Seismic Design Aids for Nonlinear Analysis of Reinforced Concrete Structures

Seismic Design Aids for Nonlinear Analysis of Reinforced Concrete Structures

Srinivasan Chandrasekaran
Luciano Nunziante
Giorgio Serino
Federico Carannante

CRC Press
Taylor & Francis Group
Boca Raton   London   New York

CRC Press is an imprint of the
Taylor & Francis Group, an **informa** business

CRC Press
Taylor & Francis Group
6000 Broken Sound Parkway NW, Suite 300
Boca Raton, FL 33487-2742

© 2010 by Taylor and Francis Group, LLC
CRC Press is an imprint of Taylor & Francis Group, an Informa business

No claim to original U.S. Government works

Printed in the United States of America on acid-free paper
10 9 8 7 6 5 4 3 2 1

International Standard Book Number: 978-1-4398-0914-3 (Hardback)

Library of Congress Cataloging-in-Publication Data

Seismic design aids for nonlinear analysis of reinforced concrete structures / authors,
   Srinivasan Chandrasekaran ... [et al.].
      p. cm.
   "A CRC title."
   Includes bibliographical references and index.
   ISBN 978-1-4398-0914-3 (hard back : alk. paper)
   1. Earthquake resistant design. 2. Reinforced concrete construction. 3. Structural
analysis (Engineering) 4. Nonlinear theories. I. Chandrasekaran, Srinivasan. II. Title.

TA658.44.S386 2010
624.1'8341--dc22   2009020201

Visit the Taylor & Francis Web site at
http://www.taylorandfrancis.com

and the CRC Press Web site at
http://www.crcpress.com

Contents

Series Preface

The Advances in Earthquake Engineering series is intended primarily for the transformation of frontier technologies and research results, as well as state-of-the-art professional practices in earthquake engineering. It will encompass various topical areas such as multidisciplinary earthquake engineering, smart structures and materials, optimal design and lifecycle cost, geotechnical engineering and soil–structure interaction, structural and system health monitoring, urban earthquake disaster mitigation, postearthquake rehabilitation and reconstruction, innovative numerical methods, as well as laboratory and field testing.

This book, *Seismic Design Aids for Nonlinear Analysis of Reinforced Concrete Structures*, serves one of the aforementioned objectives. It provides nonlinear properties of reinforced concrete elements in a comprehensive form so that practicing engineers and researchers can use them readily without solving complex equations. With the step-by-step numerical procedures presented in the book, and also through supplemental electronic material found at http://www.crcpress.com/e_products/downloads/download.asp?cat_no=K10453, the reader will find the publication a very useful and practical handbook. The book is to serve not only as a reference for graduate students in civil, structural, and construction engineering, but also as a good research directory for academicians.

Franklin Y. Cheng, PhD, PE, ASCE Distinguished Member

Editor, Advances in Earthquake Engineering Series

Series Editor

Franklin Y. Cheng, PE, honorary member of ASCE, joined the University of Missouri-Rolla as an assistant professor in 1966. In 1987, the Board of Curators of the University appointed him curators' professor; he was honored as curators' professor emeritus in 2000. He is a former senior investigator, Intelligent Systems Center, University of Missouri-Rolla. Dr. Cheng received 4 honorary professorships abroad and chaired 7 of his 24 National Science Foundation (NSF) delegations to various countries for research and development cooperation. He has also been the director of international earthquake engineering symposia and numerous state-of-the-art short courses. His work has warranted grants from several funding agencies including more than 30 from NSF. He has served as either chairman or member of 37 professional societies and committees, 12 of which are ASCE groups. He was the first chair of the Technical Administrative Committee on Analysis and Computation and initiated the Emerging Computing Technology Committee and Structural Control Committee. He also initiated and chaired the Stability Under Seismic Loading Task Group of the Structural Research Council (SSRC).

Dr. Cheng has served as a consultant for Martin Marietta Energy Systems, Inc., Los Alamos National Laboratory, and Martin & Huang International, among others. The author, coauthor, or editor of 26 books and over 250 publications, Dr. Cheng's authorship includes two textbooks, *Matrix Analysis of Structural Dynamics: Applications and Earthquake Engineering,* and *Dynamic Structural Analysis.* Dr. Cheng is the recipient of numerous honors, including the MSM-UMR Alumni Merit, ASCE State-of-the-Art twice, the Faculty Excellence, and the Halliburton Excellence awards. After receiving a BS degree (1960) from the National Cheng-Kung University, Taiwan, and a MS degree (1962) from the University of Illinois at Urbana-Champaign, he gained industrial experience with C.F. Murphy and Sargent & Lundy in Chicago, Illinois. Dr. Cheng received a PhD degree (1966) in civil engineering from the University of Wisconsin-Madison.

Preface

Seismic Design Aids for Nonlinear Analysis of Reinforced Concrete Structures (with examples and computer coding) is an attempt toward clarifying and simplifying the complexities involved in estimating some basic input parameters required for such analyses. The necessity of safe seismic design of structures is becoming a big concern for the engineering community due to the increase in damage of buildings during recent earthquakes. Most existing buildings do not comply with the current seismic codes; therefore, it is necessary to assess their structural safety and to have clear answers to questions that raise doubts about their structural safety. For most of these buildings it is necessary to prevent structural failure, although the occurrence of limited damages is usually accepted. As a matter of fact, nonlinear structural analysis has been a fundamental tool for the past 30 years, but not one widely addressed in university courses and hence not currently employed by structural engineers comfortably. On the other hand, spreading of efficient and complete computer codes of structural analysis drives them toward a passive attitude that usually opposes the full verification of the design process. While nonlinear analysis methods like static pushover are commonly accepted and recommended as a reliable tool by international codes for seismic assessment of buildings, accuracy of the estimate of seismic capacity strongly depends on input parameters of such analysis. Some of the basic inputs, namely, (1) axial force–bending moment yield interaction, (2) moment-curvature, and (3) moment-rotation characteristics accounting for appropriate nonlinearity of constitutive materials of reinforced concrete elements, need to be readdressed for an accurate pushover analysis. The design curves and tables proposed in the book are the outcome of the studies conducted by the authors using a variety of nonlinear tools, computer programs, and software. During the course of teaching, researching, and short-term courses conducted on the subject, it is felt that an appropriate use of nonlinear properties of constitutive materials is not common among design engineers using software tools. They tend to use default properties of materials as input to nonlinear analyses without realizing that a minor variation in the nonlinear characteristics of the constitutive materials like concrete and steel could result in an unsatisfactory solution leading to wrong assessment and interpretation. The main reason for such ignorance can be due to complexities involved in deriving the material properties of reinforced concrete that constitute the basic input of the nonlinear analyses.

Seismic Design Aids spans five chapters on the topics (1) axial force–bending moment yield interaction (P-M), (2) bending moment-curvature relationship (M-ϕ), (3) bending moment-rotation characteristics (M-ϑ) for beams with different support conditions and loading cases, (4) collapse multiplier of seismic loads for regular framed structures using plastic theorems, both upper bound and lower bound limit analysis theorems, and (5) verification of plastic flow rule for the developed P-M interaction domains. A detailed mathematical modeling of P-M interaction of RC rectangular beams based on international codes, namely, Italian code, Indian code, and Eurocode, currently in prevalence by defining the boundaries of the subdomains and set of analytical expressions is proposed in the first chapter. Moment-curvature relationships for beams (with no axial force) and for columns (with different levels of axial forces) are presented in Chapter 2. In Chapter 3, some practical cases of beams with relevant support conditions and loading conditions are selected for which the collapse mechanism and plastic hinge extension are presented with complete analytical expressions for moment-rotation and ductility ratios. Chapter 4 deals with determination of collapse load multipliers using plastic theorems for a few selected examples that are common cases of frames with a weak-beam, strong-column type. The developed analytical modeling of P-M interaction is verified for plastic flow rule in Chapter 5. Though the material characteristics used in *Seismic Design Aids* are limited to a few international codes, readers can easily derive the required expressions in accordance to any other international code of their choice. This is made possible by presenting the step-by-step derivation of the expressions in the relevant chapters; simply by replacing a few equations addressing the material characteristics, one can readily arrive at the desired expressions. However, using the same algorithm, the authors are certain that design engineers and researchers can easily derive other cases not addressed in this book.

We also present a step-by-step procedure to carry out pushover analysis of an example frame using the proposed design curves and tables as input parameters. Two very simple relationships are proposed for upper and lower bounds of the seismic load multiplier for regular frames of the weak-beam, strong-column type. The forecasts, shown by means of their graphical representations, qualify an optimal agreement with the relevant values obtained by pushover analysis for all the regular framed structures analyzed. Knowledge of the foreseen static multipliers, also based on an easy analytical approach, is useful both for seismic assessment and design, since the structure will be safe, by definition, under the seismic loads amplified with static *lower bounds*. The computer codes used for nonlinear optimization of collapse multiplier using static theorem and for determining kinematic multipliers are given in the additional material found on the Web site; using the program, one can easily modify the input to determine the multipliers for other cases that are not addressed in *Seismic Design Aids*. The kinematic and static multipliers for collapse loads of frames are then compared with the results obtained using the nonlinear static pushover method to show the level of confidence in the results obtained using limit analysis.

Each chapter commences with a relevant brief literature review followed by a description of the detailed mathematical modeling. Using material characteristics of concrete and steel as proposed by the codes, analytical expressions are derived, based on classical theory of nonlinear mechanics. The developed equations are followed by treatment of structural components of building frames as example problems. Tables and design curves are proposed for appropriate combinations of cross-section dimensions of beams and columns with relevant sets of percentage of tensile and compression reinforcements commonly used in design offices. *Seismic Design Aids* can be useful for capacity assessment of reinforced concrete (RC) elements whose cross-sections are known and also for performing nonlinear analysis of RC structures using readily available computer programs. Design curves are given only for few combinations of cross-section dimensions and steel reinforcement to limit the color illustrations, thereby keeping the cost affordable. Using the complementary information at http://www.crcpress.com/e_products/downloads/download .asp?cat_no=K10453 provided, one can compute the required parameters for any desired section not illustrated in the figures or tables of this book. Tables are developed in a spreadsheet form (Excel file), and steps to use these files are also described at the end of each chapter. Design engineers can readily use these tables and curves as input for their design assignments. The proposed analytical expressions of the input parameters addressed in *Seismic Design Aids* are results of extensive research work carried out by the authors. The numerical procedures are proposed in the tables after thorough verification of the results in close agreement with those obtained from analytical expressions. Complete computer coding, used for obtaining the collapse multipliers, is given at the end. With appropriate modifications in the arguments, one can easily determine the results for any specific building frame of interest.

The authors hope that *Seismic Design Aids* will be a useful reference to researchers preparing for advanced courses in structural mechanics. The authors extend their sincere thanks to the editorial board of CRC Press, Taylor & Francis Group, LLC, for publishing this book with great enthusiasm and encouragement. The authors also want to place on record the generous permission accorded by Computers and Structures Inc., Berkeley, California, for the use of screen shots of SAP2000 software in this book. The basic objective is to make nonlinear properties of RC elements available in a comprehensive form so that practicing engineers and researchers can use them readily without solving these complex equations. It is hoped that many design engineers, particularly those facing the task of seismic assessment of buildings, will find this book a very useful practical reference. We are grateful for any constructive comments or criticisms that readers wish to communicate and for notification of any errors detected in this book.

The authors have received great assistance, encouragement, and inspiration from many sources. Thanks are given to the colleagues of the Department of Structural Engineering, University of Naples Federico II, and to the Ministry of University Research (MiUR) for the fellowship assistance of one of the authors. Thanks are also given to the students of advanced courses of structural engineering and to practicing engineers who attended several training programs, workshops, and lecture series organized by the authors and their colleagues in Italy and India for giving their exciting feedback to the approach and methodology of handling the subject.

Finally, the authors would like to place on record the extensive cooperation and kindness shown by their family members during the completion of this book within the scheduled time frame.

Srinivasan Chandrasekaran
Luciano Nunziante
Giorgio Serino
Federico Carannante

About the Authors

Srinivasan Chandrasekaran is currently reader in structural engineering, Department of Civil Engineering, Institute of Technology, Banaras Hindu University, India, and also a visiting professor (MiUR Fellow) at the Department of Structural Engineering, University of Naples Federico II, Naples, Italy, under the invitation of the Ministry of Italian University Education and Research (MiUR) from 2007 to 2009. He has 17 years of experience in teaching and research, during which he has taught several basic and advanced courses in civil and structural engineering to undergraduate and postgraduate students in India and Italy. He conducted specialized research in Italy on advanced nonlinear modeling and analysis of building under seismic loads with their experimental validation. He is also a participating member of ReLuis Line 7, an executive project under the European Commission, which involves many research laboratories and universities in Europe. He conducts teaching and research in seismic engineering. He also has conducted research on nonlinear dynamic analysis of offshore compliant structures in collaboration with researchers and faculty members in Italy, India, and the United States. He has published 60 research papers in refereed journals and conferences and has successfully completed many consultancy projects in India. He also has industrial experience in design, consultancy, and execution of major civil engineering projects in India including cement plants, paper industries, and other commercial and educational institutions.

Luciano Nunziante is full professor of structural mechanics and plasticity in the Department of Structural Engineering, University of Naples Federico II, Italy, where he has taught for the last 30 years. He has authored many books on plasticity, structural mechanics, and construction science that are referred to as standard textbooks for undergraduate and postgraduate programs in civil and structural engineering in Europe and abroad. He has successfully executed many projects in cooperation with industrial partners in Italy and the European Union. He has published about 250 research papers in refereed journals and conferences. He has also organized and chaired many international conferences on advanced topics in structural mechanics.

Giorgio Serino is a full professor of structural engineering in the Department of Structural Engineering, University of Naples Federico II, Italy, where he has taught for the last 20 years. He has a rich experience that includes teaching and research in seismic analysis and design of structures, with special interest in response control of structures using passive and semiactive dampers. He is also currently the coordinator of ReLuis Line 7, an executive project under the European Commission, which involves a research team of 18 subgroups consisting of several industrial partners and universities involved in research in seismic engineering. He has authored about 100 research papers in refereed journals and conferences and successfully completed several international and national research projects and consultancy assignments.

Federico Carannante is a contract lecturer in the Department of Structural Engineering, University of Naples Federico II, Italy, and is working on the nonlinear modeling of biomechanical materials commonly used in aerospace applications. He has authored about 20 research papers in refereed journals and conferences. His special interest rests in analysis of functionally graded anisotropic materials, both elastic and nonelastic.

Disclaimer

The computer coding given in *Seismic Design Aids* is based on the hypothesis used in the analysis. In using the coding, the reader accepts and understands that no warranty is expressed or implied by the authors on the accuracy or the reliability of the programs. The examples presented are only introductory guidelines to explain the applications of the proposed methodology. The reader must explicitly understand the hypothesis made in the derivation process and must independently verify the results.

Srinivasan Chandrasekaran
Luciano Nunziante
Giorgio Serino
Federico Carannante

Naples, Italy

Notations

α_n	angle between the normal to P-M boundary and $d\varepsilon_{CG}$ axis
α_p	angle between the plastic strain vector and $d\varepsilon_{CG}$ axis
γ_c	partial safety factor for concrete
γ_s	partial safety factor for steel
δ	displacement (mm)
δ_e	elastic displacement (mm)
δ_p	plastic displacement (mm)
$\Delta\theta$	relative rotation (rad)
$\Delta\theta_E$	relative rotation at elastic limit (rad)
$\Delta\theta_u$	relative rotation at collapse (rad)
ε_c	strain in generic fiber of concrete
$\varepsilon_{c,max}$	maximum strain in concrete
ε_{c0}	elastic limit strain in concrete
ε_{cu}	ultimate limit strain in concrete
ε_{st}	strain in tensile reinforcement
ε_{sc}	strain in compression reinforcement
ε_{s0}	elastic limit strain in reinforcement
ε_{su}	ultimate limit strain in reinforcement
ε_{CG}	strain at CG of the cross-section
θ	total rotation (rad)
θ_e	total elastic rotation (rad)
θ_p	total rotation at collapse (rad)
σ_c	stress in generic fiber of concrete (N/mm^2)
$\sigma_{c,max}$	maximum stress in concrete (N/mm^2)
σ_{c0}	design ultimate stress in concrete in compression (N/mm^2)
σ_y	yield strength of steel (N/mm^2)
σ_{s0}	design ultimate stress in steel (N/mm^2)
σ_{st}	stress in tensile reinforcement (N/mm^2)
σ_{sc}	stress in compression reinforcement (N/mm^2)
ϕ	curvature (rad/m)
ϕ_e	elastic curvature (rad/m)
ϕ_E	limit elastic curvature (rad/m)
ϕ_u	ultimate curvature (rad/m)
η_θ	rotation ductility ($\Delta\theta_u/\Delta\theta_E$)
η_ϕ	curvature ductility (ϕ_u/ϕ_E)
A_{st}	area of tension reinforcement (mm^2)
A_{sc}	area of compression reinforcement (mm^2)
b	width of the beam (mm)
d	effective cover (mm)
$d\varepsilon_{CG}$	strain increment in the center of gravity (CG) of the reinforced concrete (RC) beam

$d\phi$	curvature increment
D	overall depth of the beam (mm)
E_s	modulus of elasticity in steel (N/mm^2)
k_c	collapse load multiplier
K_E^θ	rotational-elastic stiffness (kN-m/rad)
K_p^θ	rotational-hardening modulus (kN-m/rad)
K_E^ϕ	curvature-elastic stiffness (kN-m^2/rad)
K_p^ϕ	curvature-hardening modulus (kN-m^2/rad)
M	bending moment (N-m)
M_e	elastic bending moment (N-m)
M_E	limit elastic bending moment (N-m)
M_u	ultimate bending moment (N-m)
p_t	percentage of tensile reinforcement
p_c	percentage of compression reinforcement
P	axial load (N)
P_e	elastic axial load (N)
P_E	limit elastic axial load (N)
P_u	ultimate axial load (N)
q	depth of plastic kernel of concrete (mm)
R_{ck}	compressive cube strength of concrete (N/mm^2)
V	shear (kN)
x_c	depth of neutral axis measure from extreme compression fiber (mm)
$x_c^0, x_c', x_c'', x_c'''$	limit position neutral axis (mm)
y	depth of generic fiber of concrete measured from extreme compression fiber (mm)
y_Q	depth of point of rotation in subdomain 6 measured from extreme compression fiber (mm)

1 Axial Force–Bending Moment Yield Interaction

1.1 SUMMARY

The limit state design procedure of reinforced concrete elements has undergone major revision in recent times with more emphasis toward a performance-based engineering approach. This design approach demands a thorough understanding of axial force–bending moment (P-M) yield interaction of elements, for moment-resistant reinforced concrete (RC) frames under seismic loads, in particular. Current design methodologies, both recommended by international codes and employed by practicing engineers, include desirable features of ultimate strength and working stress procedures as well ensure a ductile response. In this chapter, detailed mathematical modeling of P-M yield interaction of RC rectangular beams based on Eurocode currently in prevalence is presented; six subdomains defining the boundary of P-M yield interaction are classified. A complete set of analytical expressions is proposed and also illustrated through relevant examples. Results obtained for the failure interaction curve of RC rectangular sections under P-M yield interaction show that by adopting Eurocode strain limits, the boundary curve is divided into two main parts, namely, (1) tension failure with weak reinforcement resulting in yielding of steel and (2) compression failure with strong reinforcement resulting in crushing of concrete. The curves are given in analytical form for every feasible coupling of bending moment and axial force. Advantageous use of the proposed P-M interaction for nonlinear seismic analysis is demonstrated in the subsequent chapters; also the developed boundary of different subdomains is verified for a plastic flow rule. With the help of the presented mathematical model and proposed expressions for P-M yield interaction, the designing of new structures and assessment of existing RC structures can be performed with better understanding and improved accuracy.

1.2 INTRODUCTION

Concrete is a heterogeneous, cohesive-frictional material exhibiting a complex nonlinear inelastic behavior under multiaxial stress states. The wide use of concrete as the primary structural material in several complex structures demands detailed understanding of the material response under a combination of different loads (Abu-Lebdeh and Voyiadjis 1993; Candappa, Sanjayan, and Setunge 2001; Park and Kim 2003). Sufficient ductility ensured in the design procedure is an important prerequisite for suitability of reinforced concrete structures to resist seismic loads (IS 13920, 1993); this is true because seismic design philosophy demands energy dissipation/ absorption by postelastic deformation for collapse prevention during major earthquakes (Chandrasekaran, Tripati, and Srivastav 2003; Chandrasekaran, Serino, and

1

Gupta 2008). Ductility also ensures effective redistribution of moments at critical sections as the collapse load is approached (Park and Paulay 1975; Bangash 1989; Papadrakakis, Fragiadakis, and Lagaros 2007). Ductility, a measure of energy dissipation by inelastic deformation during major earthquakes, depends mainly on the moment-curvature relationship at critical sections where plastic hinges are expected/ imposed to be formed at collapse. RC structures have the facility of changing, within certain limits, at the ultimate moment the designer pleases, without changing the overall dimensions of the cross-section. As a result, it is sometimes suggested that the reinforcement steel areas should be adjusted to make the distribution of the ultimate moment in the members the same as the elastic bending moment diagram for the factored (ultimate) load. This is a critical aspect of (intended) performance-based design of the structure, leading to some advantages, namely, (1) the elastic analysis necessary will be more laborious; (2) the resulting design shall address the required performance criteria set by the designer; as well as (3) plastic hinges are made to form on the selected structural components of the desired choice (for example, on the beam and not on the column), thus ensuring the required performance of buildings under seismic loads. In other words, the structures should be able to resist earthquakes in a quantifiable manner and to present levels of desired possible damage (Ganzerli, Pantelides, and Reaveley 2000; Ghobarah 2001). Studies (Paulay and Priestley 1992) conducted show that the behavior of statically indeterminate RC structures depends on a cross-section area of reinforcing the steel-to-concrete ratio. For smaller values of this ratio, reinforcement yields plastically before the concrete is crushed in compression, while for larger values, it may initiate crushing of concrete prior to the yielding of reinforcement. However, this ratio becomes critical when tensile steel reaches yield limit simultaneously with the extreme compressive fiber of concrete reaching its crushing strain. The increasing concern of the structural safety of existing buildings not complying with current seismic codes demands performance assessment to evaluate their seismic risk, which is a major task ahead for structural designers.

Thus, the objective of ensuring safe buildings intensifies the above-stated concerns for which pushover analysis can be seen as a rapid and reasonably accurate method (ATC-40, 1996). Pushover analysis accounts for inelastic behavior of the building models and provides a reasonable estimate of deformation capacity while identifying critical sections likely to reach limit state during earthquakes (Chopra and Goel 2000; Chao, Yungting, and Ruo 2006). Researchers used pushover analysis successfully for seismic evaluation and showed its comparison with other detailed analysis procedures (see, for example, Esra and Gulay 2005; Chandrasekaran and Roy 2004, 2006; Chandrasekaran, Nunzinate, et al. 2008b). Researchers emphasised that accuracy of results obtained from pushover analysis are strongly influenced by basic inputs like: (1) stress-strain relationship of constitutive materials; (2) P-M yield interaction; as well as (3) moment rotation capacity of members (see, for example, Chandrasekaran et al. 2008a). A qualitative insight of these inputs, P-M interaction in particular, for rectangular cross-section with different area of tensile and compressive steel accounting for nonlinear properties of constitutive materials is relatively absent in the literature. This chapter presents a mathematical development of nonlinear behavior of reinforced concrete members and derives P-M yield interaction while describing their six subdomains.

1.3 MATHEMATICAL DEVELOPMENT

Concrete under multiaxial compressive stress state exhibits significant nonlinearity, which can be successfully represented by nonlinear constitutive models (Hognestad, Hanson, and McHenry 1955; Chen and Chen 1975; Ottosen 1977; Chen 1994a, 1994b). Many researchers reported different failure criteria in stress space by a number of independent control parameters (see, for example, Hsieh, Ting, and Chen 1982; Menetrey and William 1995; Sankarasubramaniam and Rajasekaran 1996; Nunziante, Gambarotta, and Tralli 2007). A nonlinear elastic response of concrete is characterized by parabolic stress-strain relationship in the current study and shown in Figure 1.1. Elastic limit strain and strain at cracking are limited to 0.2% and 0.35%, respectively (D.M. 9 gennaio 1996). Tensile stresses in concrete are ignored in the study. The design ultimate stress in concrete in compression is given by

$$\sigma_{c0} = \frac{(0.83)(0.85)R_{ck}}{\gamma_c} \tag{1.1}$$

The stress-strain relationship for concrete under compression stresses is given by

$$\sigma_c(\varepsilon_c) = a\varepsilon_c^2 + b\varepsilon_c + c \qquad 0 < \varepsilon_c < \varepsilon_{c0}$$
$$\sigma_c(\varepsilon_c) = \varepsilon_{c0} \qquad \varepsilon_{c0} < \varepsilon_c < \varepsilon_{cu} \tag{1.2}$$

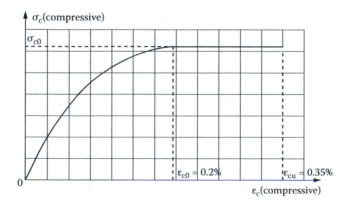

FIGURE 1.1 Stress-strain relationship for concrete.

where compression stresses and strains are assumed to be positive in the analysis. Constants a, b, and c in Equation 1.2 are determined by imposing the following conditions:

$$\sigma_c(\varepsilon_c = 0) = 0 \qquad\qquad c = 0$$

$$\sigma_c(\varepsilon_c = \varepsilon_{c0}) = \sigma_{c0} \qquad \Rightarrow \quad a\varepsilon_{c0}^2 + b\varepsilon_{c0} = \sigma_{c0}$$

$$\left[\frac{d\sigma_c}{d\varepsilon_c}\right]_{\varepsilon_c = \varepsilon_{c0}} = 0 \qquad\qquad 2a\varepsilon_{c0} + b = 0$$

$$(1.3)$$

By solving the above equations, we get

$$a = -\frac{\sigma_{c0}}{\sigma_{c0}^2}, \quad b = \frac{2\sigma_{c0}}{\sigma_{c0}}, \quad c = 0 \tag{1.4a}$$

By substituting in Equation 1.2, we get

$$\sigma_c(\varepsilon_c) = -\frac{\sigma_{c0}}{\varepsilon_{c0}^2}\varepsilon_c^2 + \frac{2\sigma_{c0}}{\varepsilon_{c0}}\varepsilon_c, \quad 0 < \varepsilon_c < \varepsilon_{c0} \tag{1.4b}$$

Steel is isotropic and homogeneous material exhibiting stress-strain relationship as shown in Figure 1.2. While the ultimate limit strain in tension and that of compression are taken as 1% and 0.35%, respectively (D.M. 9 gennaio 1996), elastic strain in steel in tension and compression are considered the same. The design ultimate stress in steel is given by

$$\sigma_{s0} = \frac{\sigma_y}{\gamma_s} \tag{1.5}$$

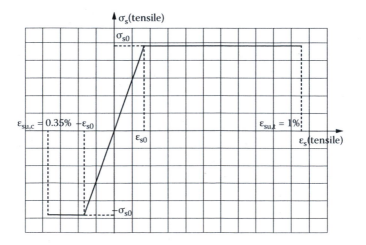

FIGURE 1.2 Stress-strain relationship for steel.

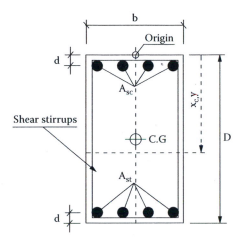

FIGURE 1.3 Cross-section of RC beam.

The stress-strain relationship for steel is given by

$$\begin{aligned}
\sigma_s(\varepsilon_s) &= E_s\varepsilon_s & -\varepsilon_{s0} &< \varepsilon_s < \varepsilon_{s0} \\
\sigma_s(\varepsilon_s) &= \sigma_{s0} & \varepsilon_{s0} &< \varepsilon_s < \varepsilon_{su,t} & (\varepsilon_{su,t} = \varepsilon_{su}) \\
\sigma_s(\varepsilon_s) &= -\sigma_{s0} & -\varepsilon_{su,c} &< \varepsilon_s < -\varepsilon_{s0}
\end{aligned} \tag{1.6}$$

The reinforced concrete beam of rectangular cross-section shown in Figure 1.3 is now examined for P-M yield interaction behavior for the different percentage of tension and compression reinforcements. The fundamental Bernoulli hypothesis of linear strain over the cross-section, both for elastic and for elastic-plastic responses, of the beam under bending moment combined with axial force is assumed. Interaction behavior becomes critical when one of the following conditions apply: (1) Reinforcement in tension steel reaches ultimate limit; (2) strain in concrete in extreme compression fiber reaches ultimate limit; or (3) maximum strain in concrete in compression reaches elastic limit under only axial compression. Figure 1.4 shows P-M limit domain consisting of six subdomains as described below. Only the upper boundary curves (corresponding to positive-bending moment M) will be examined since there exists a polar symmetry of the domains with respect to the center of the domain. Figure 1.5 shows the strain and stress profile in steel and concrete for subdomains 1 and 2 where collapse is caused by yielding of steel, whereas Figure 1.6 shows the strain and stress profile for subdomains 3 to 6 for which collapse is caused by crushing of concrete.

1.4 IDENTIFICATION OF SUBDOMAINS

1.4.1 SUBDOMAINS 1 AND 2: COLLAPSE CAUSED BY YIELDING OF STEEL

In subdomain 1 (Figures 1.4 to 1.7), the position of neutral axis measured from the origin placed on the top of rectangular section (see Figure 1.3) varies in the range]−∞, 0]. Strain in tensile steel reaches ultimate limit and the corresponding

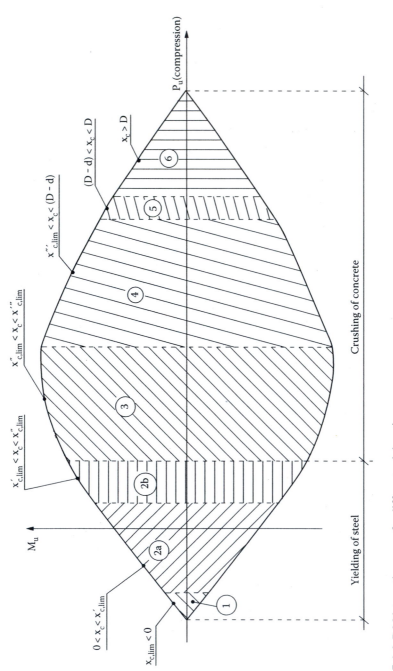

FIGURE 1.4 P-M interaction curve for different subdomains.

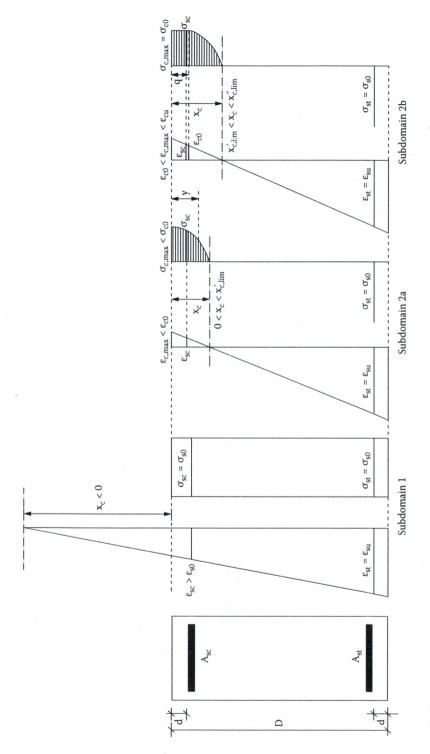

FIGURE 1.5 Collapse caused by yielding of tensile steel.

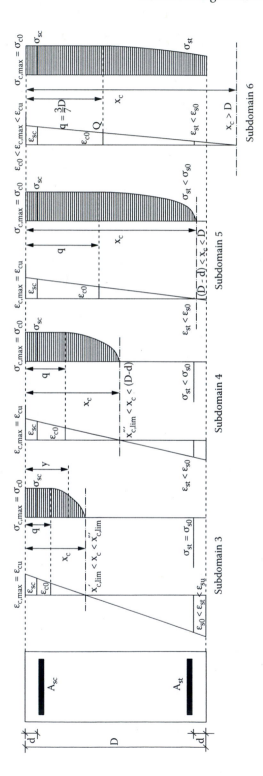

FIGURE 1.6 Collapse caused by crushing of concrete.

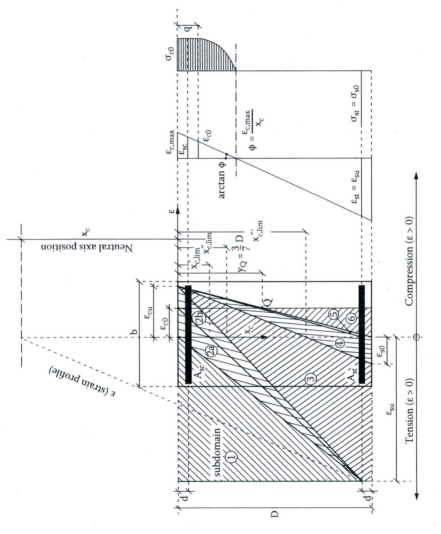

FIGURE 1.7 Strain profile for different subdomains.

stress reaches design ultimate stress, whereas strain in steel on compression zone is given by

$$\varepsilon_{sc} = \varepsilon_{su}\left(\frac{x_c - d}{D - x_c - d}\right) \quad \forall x_c < 0 \tag{1.7}$$

The strain in compression steel reaches elastic limit for the position of neutral axis assuming the value as

$$x_{c,lim}^0 = \frac{d(\varepsilon_{su} + \varepsilon_{s0}) - D\varepsilon_{s0}}{(\varepsilon_{su} - \varepsilon_{s0})} \tag{1.8}$$

for $\varepsilon_{sc} > \varepsilon_{s0}$, $\sigma_{sc} = \sigma_{s0}$, and ultimate axial force and bending moment are given by

$$\begin{cases} P_u = \sigma_{s0}(A_{sc} - A_{st}) \\ M_u = \sigma_{s0}(A_{st} + A_{sc})\left(\dfrac{D}{2} - d\right) \end{cases} \quad \forall x_c \in \left[-\infty, x_{c,lim}^0\right] \tag{1.9}$$

for $\varepsilon_{sc} \leq \varepsilon_{s0}$, $\sigma_{sc} = E_s\varepsilon_{sc}$, and ultimate axial force and bending moment (with respect to the center of gravity [CG] of the cross-section) are given by

$$\begin{cases} P_u = A_{sc}E_s\varepsilon_{su}\left(\dfrac{x_c - d}{D - x_c - d}\right) - \sigma_{s0}A_{st} \\ M_u = \left[A_{sc}E_s\varepsilon_{su}\left(\dfrac{x_c - d}{D - x_c - d}\right) + \sigma_{s0}A_{st}\right]\left(\dfrac{D}{2} - d\right) \end{cases} \quad \forall x_c \in \left[x_{c,lim}^0, 0\right] \tag{1.10}$$

From the above equations, position of neutral axis can be deduced as

$$x_c = D - d - \frac{A_{sc}E_s\varepsilon_{su}(D - 2d)}{Pu + A_{sc}E_s\varepsilon_{su} + A_{st}\sigma_{s0}} \tag{1.11}$$

By substituting in Equation 1.10, we get

$$M_u = \left(\frac{D}{2} - d\right)(P_u + 2A_{st}\sigma_{s0}) \tag{1.12}$$

For depth of neutral axis becoming zero, ultimate axial force and bending moment are given by

$$\begin{cases} P_u = -A_{sc}E_s\varepsilon_{su}\left(\dfrac{d}{D - d}\right) - \sigma_{s0}A_{st} \\ M_u = \left[\sigma_{s0}A_{st} - A_{sc}E_s\varepsilon_{su}\left(\dfrac{d}{D - d}\right)\right]\left(\dfrac{D}{2} - d\right) \end{cases} \quad \text{for } x_c = 0 \tag{1.13}$$

Subdomain 2 is further composed of two regions, namely, yielding of tensile steel while strain in concrete remains within elastic limits (2a), and yielding of tensile steel while strain in concrete reaches ultimate limit (2b). Depth of neutral axis in these regions lies in the range $[0, x'_{c,lim}]$ and $[x'_{c,lim}, x''_{c,lim}]$ for regions (2a) and (2b), respectively (Figure 1.7), and are given by

$$x'_{c,lim} = \left(\frac{\varepsilon_{c0}}{\varepsilon_{c0} + \varepsilon_{cu}} \right)(D-d) = 0.167(D-d) \quad \text{for} \quad \varepsilon_{c,max} = \varepsilon_{c0}; \quad \varepsilon_{st} = \varepsilon_{su} \quad (1.14)$$

$$x''_{c,lim} = \left(\frac{\varepsilon_{cu}}{\varepsilon_{su} + \varepsilon_{cu}} \right)(D-d) = 0.259(D-d) \quad \text{for} \quad \varepsilon_{c,max} = \varepsilon_{cu}; \quad \varepsilon_{st} = \varepsilon_{su} \quad (1.15)$$

Strain in compression steel, in subdomain 2a, is given by

$$\varepsilon_{sc} = \varepsilon_{su} \left(\frac{x_c - d}{D - x_c - d} \right) \quad \forall x_c \in [0, x'_{c,lim}] \quad (1.16)$$

The stress and strain of a generic compression fiber of concrete located at a distance y measured from the extreme compression fiber are given by

$$\sigma_c(\varepsilon_c(y)) = -\frac{\sigma_{c0}}{\varepsilon_{c0}^2}\varepsilon_c^2 + \frac{2\sigma_{c0}}{\varepsilon_{c0}}\varepsilon_c = \frac{(x_c - y)[2\varepsilon_{c0}(D - x_c - d) + \varepsilon_{su}(y - x_c)]\sigma_{c0}\varepsilon_{su}}{\varepsilon_{c0}^2(x_c + d - D)^2} \quad (1.17)$$

$$\varepsilon_c = \varepsilon_{su} \left(\frac{x_c - y}{D - x_c - d} \right) \quad \forall x_c \in [0, x'_{c,lim}] \quad (1.18)$$

Ultimate axial force and bending moment in subdomain 2a are given by

$$\begin{cases} P_u = \int_0^{x_c} b\sigma_c(\varepsilon_c(y))dy - \sigma_{s0}A_{st} + \sigma_{sc}A_{sc} \\[4mm] M_u = \int_0^{x_c} b\sigma_c(\varepsilon_c(y))\left(\frac{D}{2} - y\right)dy + (\sigma_{s0}A_{st} + \sigma_{sc}A_{sc})\left(\frac{D}{2} - d\right) \end{cases} \quad 0 < x_c \le x'_{c,lim} \quad (1.19)$$

Depth of *plastic kernel* of concrete is given by

$$q = x_c - \frac{\varepsilon_{c0}}{\varepsilon_{su}}(D - x_c - d) \quad (1.20)$$

Ultimate axial force and bending moment in subdomain 2b are given by

$$
\left\{
\begin{aligned}
P_u &= \int_q^{x_c} b\sigma_c(\varepsilon_c(y))\,dy + bq\sigma_{c0} - \sigma_{s0}A_{st} + \sigma_{sc}A_{sc} \qquad x'_{c,\lim} < x_c \le x''_{c,\lim} \\[2mm]
M_u &= \int_q^{x_c} b\sigma_c(\varepsilon_c(y))\left(\frac{D}{2}-y\right)dy + \frac{bq\sigma_{c0}}{2}(D-q) + (\sigma_{s0}A_{st} + \sigma_{sc}A_{sc})\left(\frac{D}{2}-d\right)
\end{aligned}
\right.
$$

$$(1.21)$$

1.4.2 SUBDOMAINS 3 TO 6: COLLAPSE CAUSED BY CRUSHING OF CONCRETE

In subdomain 3, collapse occurs when maximum strain in concrete reaches crushing strain while strain in tension steel varies in the range $[\varepsilon_{s0},\varepsilon_{su}]$, stress in tensile steel is σ_{s0}, and position of neutral axis varies in the range $[x''_{c,\lim},x'''_{c,\lim}]$ (Figures 1.4 to 1.7). Position of neutral axis $x'''_{c,\lim}$, characterized by $\varepsilon_{st} = \varepsilon_{s0}$, is given by

$$
x'''_{c,\lim} = \frac{\varepsilon_{cu}}{\varepsilon_{s0}+\varepsilon_{cu}}(D-d)
$$

$$(1.22)$$

Strains in steel, both in tension and compression, are given by

$$
\varepsilon_{st} = \varepsilon_{cu}\left(\frac{D-x_c-d}{x_c}\right), \quad \varepsilon_{sc} = \varepsilon_{cu}\left(\frac{x_c-d}{x_c}\right) \quad \forall x_c \in [x''_{c,\lim},x'''_{c,\lim}]
$$

$$(1.23)$$

The corresponding stress in steel bars reaches ultimate limit as strain exceeds elastic limit. Strain in the generic fiber of concrete and depth of plastic kernel are given by

$$
\varepsilon_c = \varepsilon_{cu}\left(\frac{x_c-y}{x_c}\right), \quad q = \frac{\varepsilon_{cu}-\varepsilon_{c0}}{\varepsilon_{cu}}x_c \quad \forall x_c \in [x''_{c,\lim},x'''_{c,\lim}]
$$

$$(1.24)$$

In subdomains 4 and 5, the position of the neutral axis varies in the range $[x'''_{c,\lim},(D-d)]$ and $[(D-d),D]$, respectively. In subdomain 4, strain in tensile steel varies in the range $[0,\varepsilon_{s0}]$, stress in tensile steel is $\sigma_{st} = E_s\varepsilon_{st}$, whereas in subdomain 5, tensile steel gains compressive stress progressively. Ultimate axial force and bending moment in subdomains 3 to 5 are given by

$$
\left\{
\begin{aligned}
P_u &= \int_q^{x_c} b\sigma_c(\varepsilon_c(y))\,dy + bq\sigma_{c0} - \sigma_{st}A_{st} + \sigma_{sc}A_{sc} \\[2mm]
M_u &= \int_q^{x_c} b\sigma_c(\varepsilon_c(y))\left(\frac{D}{2}-y\right)dy + \frac{bq\sigma_{c0}}{2}(D-q) + (\sigma_{st}A_{st} + \sigma_{sc}A_{sc})\left(\frac{D}{2}-d\right)
\end{aligned}
\right.
$$

$$(1.25)$$

Figure 1.7 shows the linear strain profile over the cross-section determined by the tensile strain in steel. In subdomain 6, the position of the neutral axis varies in the range $[D, +\infty[$, and strain diagram in cross-section rotates about point Q as shown in Figure 1.7. The depth of plastic kernel, whose distance is measured from the extreme compression fiber, is given by

$$q = y_Q = \left(\frac{\varepsilon_{cu} - \varepsilon_{c0}}{\varepsilon_{cu}} \right) D = \frac{3}{7} D \tag{1.26}$$

Strain in reinforcing steel, both in tension and compression, are given by

$$\varepsilon_{st} = \frac{\varepsilon_{cu} \varepsilon_{c0} (D - x_c - d)}{\varepsilon_{cu} x_c - D(\varepsilon_{cu} - \varepsilon_{c0})}, \quad \varepsilon_{sc} = \frac{\varepsilon_{cu} \varepsilon_{c0} (x_c - d)}{\varepsilon_{cu} x_c - D(\varepsilon_{cu} - \varepsilon_{c0})} \tag{1.27}$$

The strain in generic fiber of the concrete is given by

$$\varepsilon_c = \frac{\varepsilon_{cu} \varepsilon_{c0} (x_c - y)}{\varepsilon_{cu} x_c - D(\varepsilon_{cu} - \varepsilon_{c0})} \tag{1.28}$$

Ultimate axial force and bending moment are given by

$$\begin{cases} P_u = \int_q^D b \sigma_c(\varepsilon_c(y)) dy + q \, b \sigma_{c0} - \sigma_{st} A_{st} + \sigma_{sc} A_{sc} \\ \\ M_u = \int_q^D b \sigma_c(\varepsilon_c(y)) \left(\frac{D}{2} - y \right) dy + \frac{bq\sigma_{c0}}{2} (D - q) + (\sigma_{st} A_{st} + \sigma_{sc} A_{sc}) \left(\frac{D}{2} - d \right) \end{cases} \quad D \le x_c < +\infty \tag{1.29}$$

1.5 NUMERICAL STUDIES AND DISCUSSIONS

Using the above expressions, P-M yield interaction is now studied for RC beams of different cross-sections, reinforced in both tension and compression zones. The cross-section dimensions and other relevant data can be seen from the legend of the figures. All six subdomains are traced and plotted as seen in Figures 1.8 to 1.19. The sample plots are shown for relevant practical cases, namely, (1) for different cross-sections; (2) for varying percentage in tension and compression reinforcements; (3) for different characteristic strength of concrete; and (4) for different yield strength

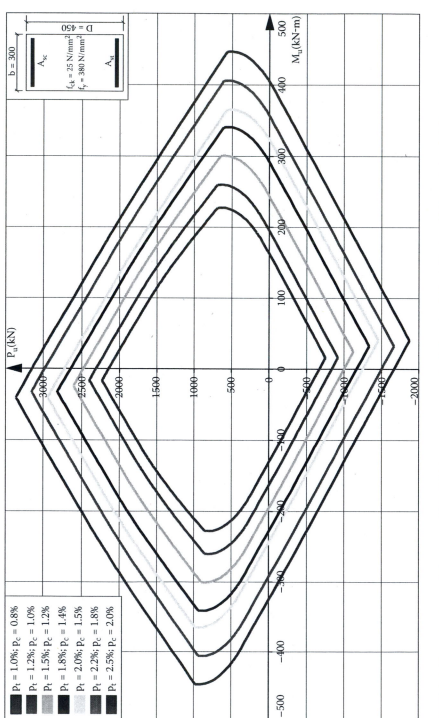

FIGURE 1.8 (See color insert following p. 138.) P-M interaction curves for RC section 300×450 ($f_{ck} = 25$ N/mm^2, $f_y = 380$ N/mm^2).

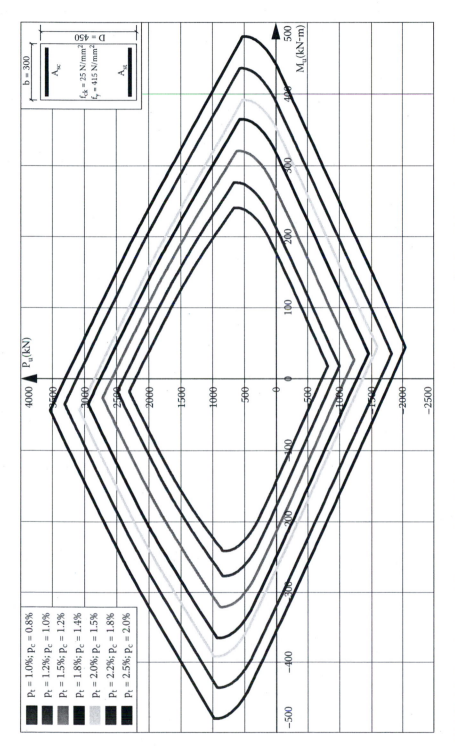

FIGURE 1.9 (See color insert following p. 138.) P-M interaction curves for RC section 300×450 ($f_{ck} = 25$ N/mm², $f_y = 415$ N/mm²).

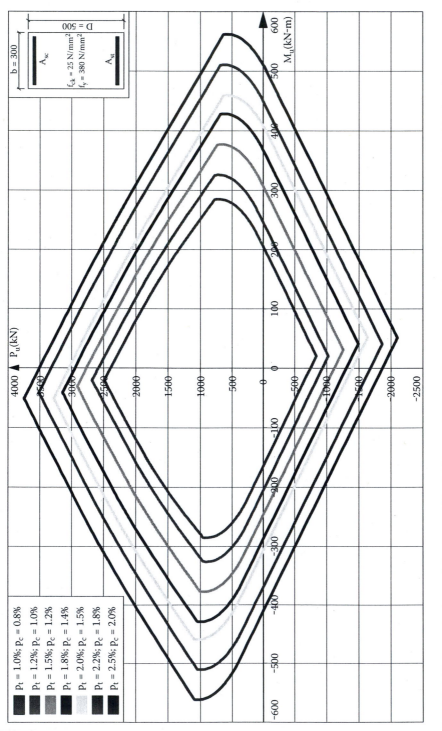

FIGURE 1.10 (See color insert following p. 138.) P-M interaction curves for RC section 300×500 ($f_{ck} = 25$ N/mm^2, $f_y = 380$ N/mm^2).

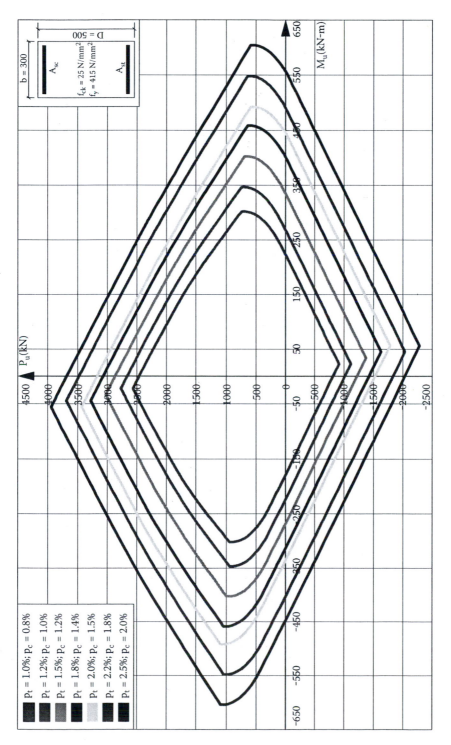

FIGURE 1.11 (See color insert following p. 138.) P-M interaction curves for RC section 300×500 ($f_{ck} = 25$ N/mm², $f_y = 415$ N/mm².

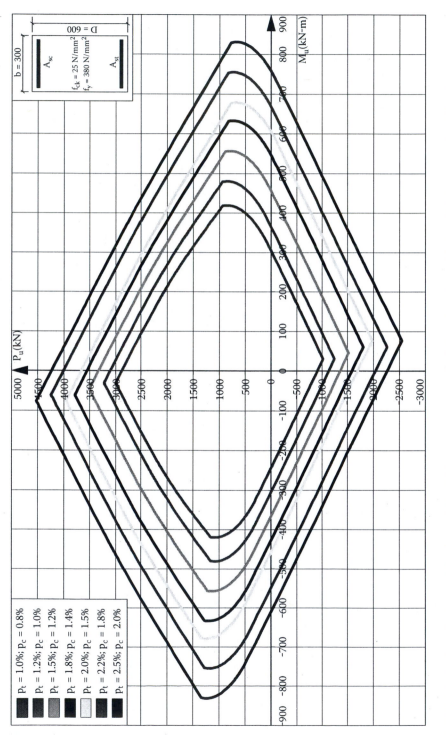

FIGURE 1.12 (See color insert following p. 138.) P-M interaction curves for RC section 300 × 600 (f_{ck} = 25 N/mm², f_y = 380 N/mm²).

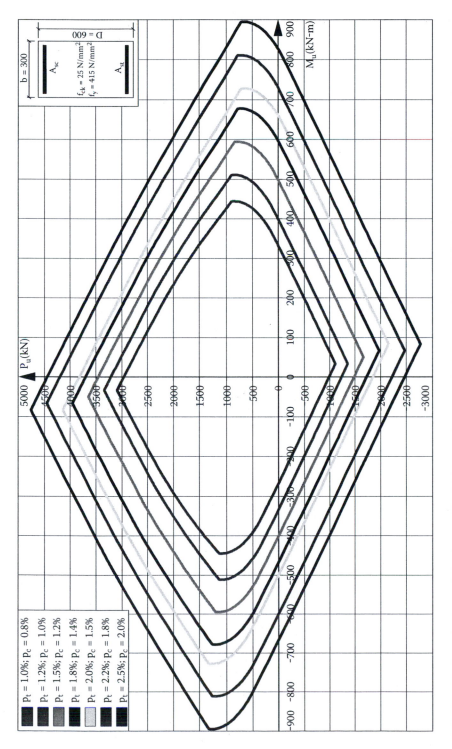

FIGURE 1.13 (See color insert following p. 138.) P-M interaction curves for RC section 300×600 ($f_{ck} = 25$ N/mm², $f_y = 415$ N/mm²).

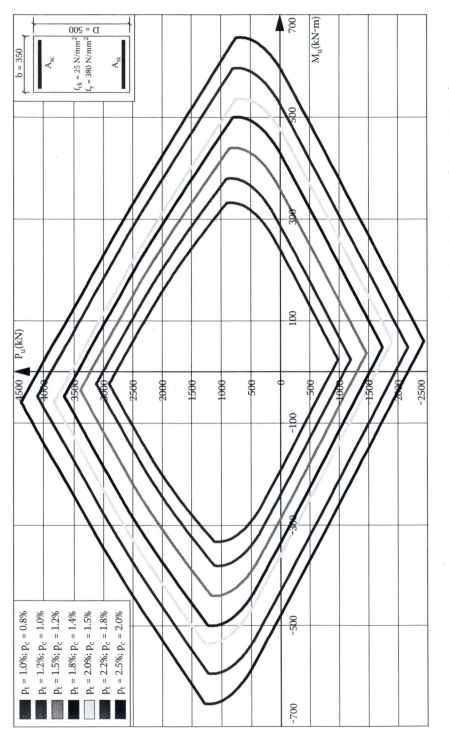

FIGURE 1.14 (See color insert following p. 138.) P-M interaction curves for RC section 350×500 ($f_{ck} = 25$ N/mm^2, $f_y = 380$ N/mm^2).

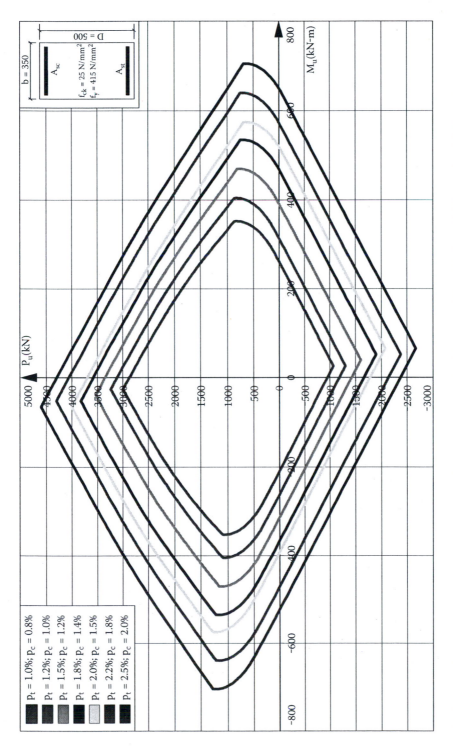

FIGURE 1.15 (See color insert following p. 138.) P–M interaction curves for RC section 350×500 ($f_{ck} = 25$ N/mm², $f_y = 415$ N/mm²).

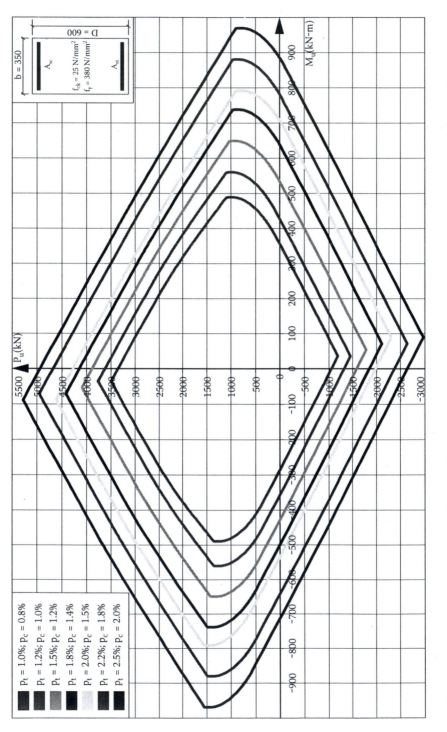

FIGURE 1.16 (See color insert following p. 138.) P-M interaction curves for RC section 350×600 ($f_{ck} = 25$ N/mm², $f_y = 380$ N/mm²).

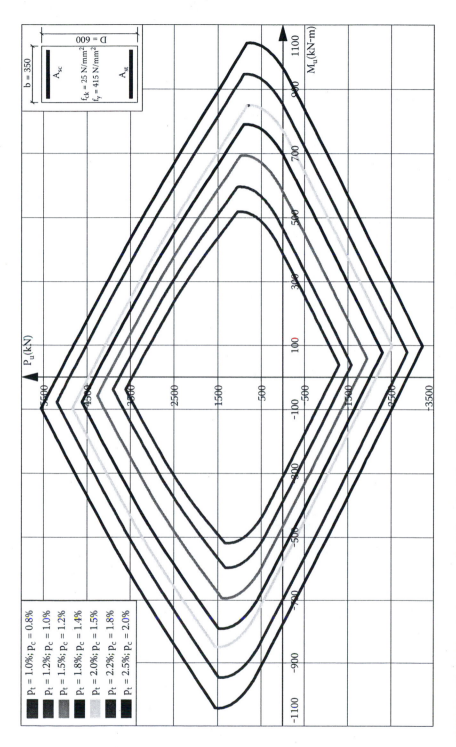

FIGURE 1.17 P-M interaction curves for RC section 350×600 ($f_{ck} = 25$ N/mm^2, $f_y = 415$ N/mm^2).

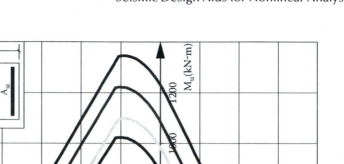

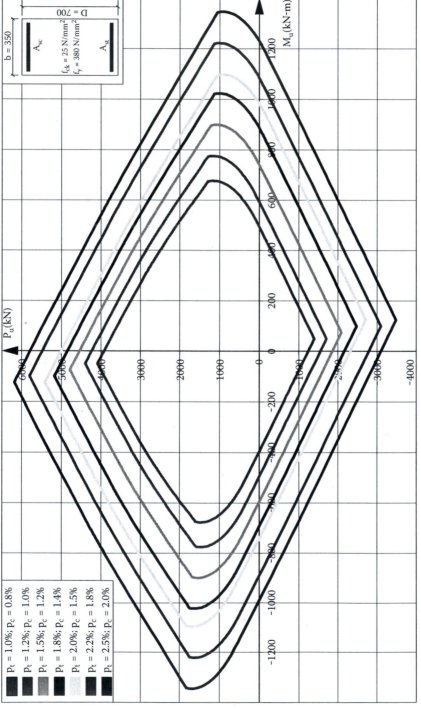

FIGURE 1.18 P-M interaction curves for RC section 350 × 700 (f_{ck} = 25 N/mm², f_y = 380 N/mm²).

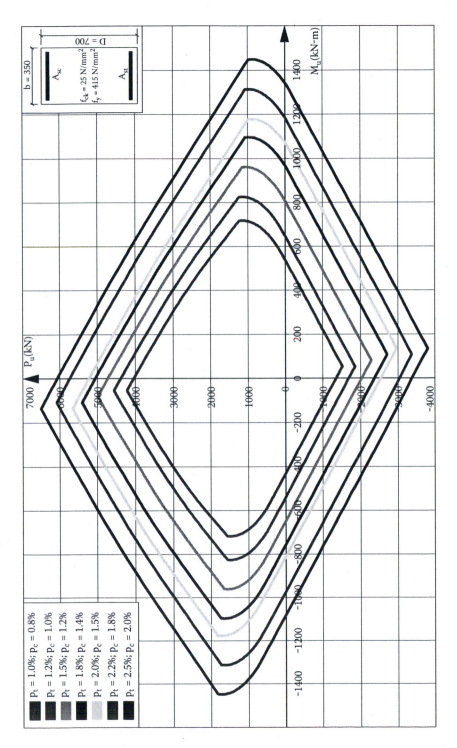

FIGURE 1.19 P-M interaction curves for RC section 350×700 ($f_{ck} = 25$ N/mm^2, $f_y = 415$ N/mm^2).

of reinforcing steel. The relevant numerical values are also reported in Tables 1.1 to 1.12. For a ready use of obtaining P-M interaction of any desired RC section other than those given in the figures and tables presented above, a summary of expressions in the closed form is given in Table 1.13. Ready use of the above-presented procedure is demonstrated using the simple spreadsheet program, as discussed in Section 1.7.

The results obtained for the RC failure interaction curve of beams of rectangular cross-section under P-M yield interaction show that, by adopting Eurocode strain limits, the boundary curve is first divided into two parts based on the type of failure, namely, (1) tension failure with weak reinforcement resulting in yielding of steel and (2) compression failure with strong reinforcement resulting in crushing of concrete. The expressions for different subdomains are also given in analytical form for every feasible coupling of bending and axial force. The boundary curve for the steel failure, in which by definition the tensile steel is in ultimate yielding condition, can be further subdivided in three subdomains (1, 2a, 2b). These parts, when subjected to increasing compressive axial force, correspond to compression of concrete reaching the ultimate limit. Subsequently, for the concrete failure part, for which by definition concrete is crushed, the curve can be subdivided into four subdomains (3, 4, 5, 6) for which by increasing the compressive axial force, strain in steel varies between the tensile failure limit and the tensile elastic limit, until elastic limit in compression for concrete disperses all over the cross-section. The sharp bend seen in the boundary of subdomain 3 to 4 corresponds to the fact that the stress-strain relationship for steel is bilinear (Figure 1.2). The procedure proposed for the bending moment-curvature relationships of beams, in the presence of constant axial force, is very simple and furnishes expressions in closed form for elastic-plastic regions of RC sections after verification with the numerical results. The applied principle may not be new, but the presented expressions in closed form will be very useful for practical nonlinear static analyses like pushover. The analyses of these responses show that for every kind of reinforcement, owing to the small bending strength increment between the elastic and the failure limits, the moment-curvature response of rectangular RC sections is basically bilinear: After the first linear response, a further linear plastic branch is present, with a small slope. The whole response is very close to an elastic-plastic response, characterized by a sort of small "hardening effect."

It is important to note that the subdomains classified for P-M interaction are based on strain limit conditions imposed by the codes. For these prescribed strain limits, all points of the section are not in failure condition; this implies that the stress and strain increments of the points lying along the P-M boundary depend on both the elastic and plastic increments. Hence, normality rule that is valid for true plastic domain does not hold completely true. Detailed discussion of the validity and applicability of flow rule can be seen in Chapter 5. Subsequent verifications made by the authors show that plastic flow rule is completely satisfied in the subdomains where failure is caused by yielding of steel, and not verified for subdomains where failure is caused by crushing of concrete.

TABLE 1.1
P-M Interaction Values for RC Section 300×450 ($p_t = 1.5\%$, $p_c = 1.2\%$, $f_{ck} = 25$ N/mm^2, $f_y = 380$ N/mm^2)

x_c	$\varepsilon_{c,max}$	$\sigma_{c,max}$	ε_{st}	σ_{st}	ε_{sc}	σ_{sc}	P_u	M_u+	M_u-
-1000	0.00000	0	0.01000	330435	-0.01000	-330435	-1124.14	24.36	24.36
-3.500	0.00000	0	0.01000	330435	-0.00901	-330435	-1124.14	24.36	24.36
-1.750	0.00000	0	0.01000	330435	-0.00820	-330435	-1124.14	24.36	24.36
0.000	0.00000	0	0.01000	330435	-0.00071	-150000	-851.32	77.56	-42.14
0.021	0.00053	5038	0.01000	330435	-0.00023	-47368	-679.47	111.45	-83.60
0.046	0.00121	9325	0.01000	330435	0.00041	86916	-420.21	162.63	-144.68
0.070	0.00200	11023	0.01000	330435	0.00114	240000	-107.31	223.22	-216.55
0.084	0.00249	11023	0.01000	330435	0.00159	330435	77.44	258.13	-258.13
0.099	0.00309	11023	0.01000	330435	0.00216	330435	132.31	266.70	-266.70
0.109	0.00350	11023	0.01000	330435	0.00254	330435	166.60	271.59	-271.59
0.172	0.00350	11023	0.00504	330435	0.00289	330435	336.06	289.91	-289.91
0.235	0.00350	11023	0.00274	330435	0.00305	330435	505.52	299.30	-299.30
0.290	0.00350	11023	0.00157	330435	0.00314	330435	650.77	300.25	-300.25
0.322	0.00350	11023	0.00106	222786	0.00317	330435	941.40	257.99	-265.93
0.374	0.00350	11023	0.00043	89499	0.00322	330435	1332.80	199.83	-217.59
0.420	0.00350	11023	0.00000	0	0.00325	330435	1624.01	153.98	-178.33
0.429	0.00350	11023	-0.00007	-15420	0.00326	330435	1677.25	145.20	-170.70
0.440	0.00350	11023	-0.00016	-32611	0.00326	330435	1737.85	135.04	-161.80
0.450	0.00350	11023	-0.00023	-49000	0.00327	330435	1796.93	124.92	-152.89
1.450	0.00231	11023	-0.00164	-330435	0.00226	330435	2600.44	-22.45	-26.26
5.000	0.00208	11023	-0.00191	-330435	0.00207	330435	2611.49	-24.23	-24.49

TABLE 1.2
P-M Interaction Values for RC Section 300 × 450 (p_t = 1.5%, p_c = 1.2%, f_{ck} = 25 N/mm², f_y = 415 N/mm²)

x_c	$\varepsilon_{c,max}$	$\sigma_{c,max}$	ε_{st}	σ_{st}	ε_{sc}	σ_{sc}	P_u	M_u+	M_u-
-1000	0.00000	0	0.01000	360870	-0.01000	-360870	-1227.68	26.60	26.60
-3.500	0.00000	0	0.01000	360870	-0.00901	-360870	-1227.68	26.60	26.60
-1.750	0.00000	0	0.01000	360870	-0.00820	-360870	-1227.68	26.60	26.60
0.000	0.00000	0	0.01000	360870	-0.00071	-150000	-908.84	88.77	-51.12
0.021	0.00053	5038	0.01000	360870	-0.00023	-47368	-736.99	122.66	-92.57
0.046	0.00121	9325	0.01000	360870	0.00041	86916	-477.73	173.85	-153.66
0.070	0.00200	11023	0.01000	360870	0.00114	240000	-164.84	234.43	-225.52
0.084	0.00249	11023	0.01000	360870	0.00159	334682	26.34	270.60	-268.67
0.099	0.00309	11023	0.01000	360870	0.00216	360870	120.80	286.89	-286.89
0.109	0.00350	11023	0.01000	360870	0.00254	360870	155.10	291.78	-291.78
0.169	0.00350	11023	0.00518	360870	0.00288	360870	317.02	309.47	-309.47
0.230	0.00350	11023	0.00290	360870	0.00304	360870	478.93	319.02	-319.02
0.282	0.00350	11023	0.00172	360870	0.00313	360870	617.72	320.71	-320.71
0.316	0.00350	11023	0.00115	241063	0.00317	360870	936.72	274.36	-283.19
0.372	0.00350	11023	0.00046	95748	0.00322	360870	1359.47	211.75	-231.29
0.420	0.00350	11023	0.00000	0	0.00325	360870	1670.03	162.95	-189.55
0.429	0.00350	11023	-0.00007	-15420	0.00326	360870	1723.26	154.18	-181.91
0.440	0.00350	11023	-0.00016	-32611	0.00326	360870	1783.86	144.01	-173.02
0.450	0.00350	11023	-0.00023	-49000	0.00327	360870	1842.95	133.90	-164.11
1.450	0.00231	11023	-0.00164	-344114	0.00226	360870	2672.31	-18.52	-33.45
5.000	0.00208	11023	-0.00191	-360870	0.00207	360870	2715.03	-26.47	-26.73

TABLE 1.3

P-M Interaction Values for RC Section 300 × 500 (p_t = 1.5%, p_c = 1.2%, f_{ck} = 25 N/mm², f_y = 380 N/mm²)

x_c	$\varepsilon_{c,max}$	$\sigma_{c,max}$	ε_{st}	σ_{st}	ε_{sc}	σ_{sc}	P_u	M_u+	M_u-
-1000	0.00000	0	0.01000	330435	-0.01000	-330435	-1257.97	30.75	30.75
-3.500	0.00000	0	0.01000	330435	-0.00889	-330435	-1257.97	30.75	30.75
-1.750	0.00000	0	0.01000	330435	-0.00802	-330435	-1257.97	30.75	30.75
0.000	0.00000	0	0.01000	330435	-0.00064	-134043	-925.67	103.86	-60.63
0.024	0.00053	5038	0.01000	330435	-0.00015	-30571	-731.94	146.89	-113.29
0.051	0.00121	9325	0.01000	330435	0.00050	104812	-439.95	211.69	-190.69
0.078	0.00200	11023	0.01000	330435	0.00123	259149	-87.69	288.32	-281.69
0.094	0.00249	11023	0.01000	330435	0.00169	330435	86.66	325.09	-325.09
0.111	0.00309	11023	0.01000	330435	0.00226	330435	148.06	335.71	-335.71
0.122	0.00350	11023	0.01000	330435	0.00264	330435	186.44	341.77	-341.77
0.193	0.00350	11023	0.00504	330435	0.00296	330435	376.07	364.37	-364.37
0.264	0.00350	11023	0.00274	330435	0.00310	330435	565.70	375.79	-375.79
0.324	0.00350	11023	0.00157	330435	0.00318	330435	728.24	376.69	-376.69
0.361	0.00350	11023	0.00106	222786	0.00321	330435	1053.47	323.19	-333.21
0.419	0.00350	11023	0.00043	89499	0.00325	330435	1491.47	249.57	-272.00
0.470	0.00350	11023	0.00000	0	0.00328	330435	1817.34	191.57	-222.32
0.479	0.00350	11023	-0.00007	-13810	0.00328	330435	1870.64	181.66	-213.69
0.490	0.00350	11023	-0.00014	-29280	0.00329	330435	1931.47	170.16	-203.64
0.500	0.00350	11023	-0.00021	-44100	0.00329	330435	1990.93	158.72	-193.58
1.500	0.00233	11023	-0.00160	-330435	0.00229	330435	2895.93	-27.97	-33.53
5.000	0.00209	11023	-0.00189	-330435	0.00208	330435	2910.36	-30.55	-30.95

TABLE 1.4

P-M Interaction Values for RC Section 300 × 500 ($p_t = 1.5\%$, $p_c = 1.2\%$, $f_{ck} = 25$ N/mm^2, $f_y = 415$ N/mm^2)

x_c	$\varepsilon_{c,max}$	$\sigma_{c,max}$	ε_{st}	σ_{st}	ε_{sc}	σ_{sc}	P_u	M_u+	M_u-
−1000	0.00000	0	0.01000	360870	−0.01000	−360870	−1373.83	33.58	33.58
−3.500	0.00000	0	0.01000	360870	−0.00889	−360870	−1373.83	33.58	33.58
−1.750	0.00000	0	0.01000	360870	−0.00802	−360870	−1373.83	33.58	33.58
0.000	0.00000	0	0.01000	360870	−0.00064	−134043	−990.04	118.02	−71.96
0.024	0.00053	5038	0.01000	360870	−0.00015	−30571	−796.31	161.05	−124.62
0.051	0.00121	9325	0.01000	360870	0.00050	104812	−504.32	225.85	−202.02
0.078	0.00200	11023	0.01000	360870	0.00123	259149	−152.06	302.48	−293.01
0.094	0.00249	11023	0.01000	360870	0.00169	354606	63.18	348.25	−347.66
0.111	0.00309	11023	0.01000	360870	0.00226	360870	135.19	361.20	−361.20
0.122	0.00350	11023	0.01000	360870	0.00264	360870	173.56	367.26	−367.26
0.190	0.00350	11023	0.00518	360870	0.00295	360870	354.76	389.09	−389.09
0.257	0.00350	11023	0.00290	360870	0.00309	360870	535.95	400.72	−400.72
0.315	0.00350	11023	0.00172	360870	0.00317	360870	691.26	402.56	−402.56
0.354	0.00350	11023	0.00115	241063	0.00320	360870	1048.23	343.88	−355.03
0.416	0.00350	11023	0.00046	95748	0.00325	360870	1521.31	264.63	−289.30
0.470	0.00350	11023	0.00000	0	0.00328	360870	1868.84	202.90	−236.48
0.479	0.00350	11023	−0.00007	−13810	0.00328	360870	1922.14	192.99	−227.85
0.490	0.00350	11023	−0.00014	−29280	0.00329	360870	1982.97	181.49	−217.80
0.500	0.00350	11023	−0.00021	−44100	0.00329	360870	2042.42	170.05	−207.74
1.500	0.00233	11023	−0.00160	−336467	0.00229	360870	2960.18	−19.45	−45.44
5.000	0.00209	11023	−0.00189	−360870	0.00208	360870	3026.22	−33.38	−33.78

TABLE 1.5

P-M Interaction Values for RC Section 300 × 600 (p_t = 1.5%, p_c = 1.2%, f_{ck} = 25 N/mm^2, f_y = 380 N/mm^2)

x_c	$\varepsilon_{c,max}$	$\sigma_{c,max}$	ε_{st}	σ_{st}	ε_{sc}	σ_{sc}	P_u	M_u+	M_u-
-1000	0.00000	0	0.01000	330435	-0.00999	-330435	-1525.62	45.77	45.77
-3.500	0.00000	0	0.01000	330435	-0.00867	-330435	-1525.62	45.77	45.77
-1.750	0.00000	0	0.01000	330435	-0.00767	-330435	-1525.62	45.77	45.77
0.000	0.00000	0	0.01000	330435	-0.00053	-110526	-1074.37	167.61	-106.53
0.029	0.00053	5038	0.01000	330435	-0.00003	-5817	-836.88	232.19	-185.61
0.062	0.00121	9325	0.01000	330435	0.00062	131185	-479.44	329.04	-301.44
0.095	0.00200	11023	0.01000	330435	0.00137	287368	-48.44	443.43	-437.46
0.113	0.00249	11023	0.01000	330435	0.00183	330435	105.09	482.13	-482.13
0.135	0.00309	11023	0.01000	330435	0.00240	330435	179.56	497.52	-497.52
0.148	0.00350	11023	0.01000	330435	0.00279	330435	226.11	506.28	-506.28
0.234	0.00350	11023	0.00504	330435	0.00305	330435	456.08	538.79	-538.79
0.320	0.00350	11023	0.00274	330435	0.00317	330435	686.06	554.85	-554.85
0.393	0.00350	11023	0.00157	330435	0.00323	330435	883.18	555.54	-555.54
0.437	0.00350	11023	0.00106	222786	0.00326	330435	1277.62	475.60	-490.51
0.508	0.00350	11023	0.00043	89499	0.00329	330435	1808.80	365.63	-399.00
0.570	0.00350	11023	0.00000	0	0.00332	330435	2204.01	279.06	-324.82
0.579	0.00350	11023	-0.00005	-11425	0.00332	330435	2257.41	266.86	-314.21
0.590	0.00350	11023	-0.00012	-24313	0.00332	330435	2318.58	252.70	-301.84
0.600	0.00350	11023	-0.00018	-36750	0.00333	330435	2378.59	238.61	-289.47
1.600	0.00238	11023	-0.00153	-322149	0.00234	330435	3463.95	-34.75	-55.64
5.000	0.00211	11023	-0.00187	-330435	0.00210	330435	3507.86	-45.35	-46.19

TABLE 1.6

P-M Interaction Values for RC Section 300 × 600 (p_t = 1.5%, p_c = 1.2%, f_{ck} = 25 N/mm², f_y = 415 N/mm²)

x_c	$\varepsilon_{c,max}$	$\sigma_{c,max}$	ε_{st}	σ_{st}	ε_{sc}	σ_{sc}	P_u	M_u+	M_u-
−1000	0.00000	0	0.01000	360870	−0.00999	−360870	−1666.13	49.98	49.98
−3.500	0.00000	0	0.01000	360870	−0.00867	−360870	−1666.13	49.98	49.98
−1.750	0.00000	0	0.01000	360870	−0.00767	−360870	−1666.13	49.98	49.98
0.000	0.00000	0	0.01000	360870	−0.00053	−110526	−1152.43	188.68	−123.39
0.029	0.00053	5038	0.01000	360870	−0.00003	−5817	−914.94	253.27	−202.48
0.062	0.00121	9325	0.01000	360870	0.00062	131185	−557.51	350.12	−318.30
0.095	0.00200	11023	0.01000	360870	0.00137	287368	−126.51	464.51	−454.33
0.113	0.00249	11023	0.01000	360870	0.00183	360870	89.48	520.07	−520.07
0.135	0.00309	11023	0.01000	360870	0.00240	360870	163.95	535.46	−535.46
0.148	0.00350	11023	0.01000	360870	0.00279	360870	210.49	544.22	−544.22
0.230	0.00350	11023	0.00518	360870	0.00304	360870	430.24	575.63	−575.63
0.312	0.00350	11023	0.00290	360870	0.00316	360870	649.98	592.03	−592.03
0.382	0.00350	11023	0.00172	360870	0.00323	360870	838.33	594.14	−594.14
0.429	0.00350	11023	0.00115	241063	0.00326	360870	1271.26	506.45	−523.04
0.504	0.00350	11023	0.00046	95748	0.00329	360870	1845.00	388.06	−424.78
0.570	0.00350	11023	0.00000	0	0.00332	360870	2266.46	295.92	−345.90
0.579	0.00350	11023	−0.00005	−11425	0.00332	360870	2319.86	283.72	−335.28
0.590	0.00350	11023	−0.00012	−24313	0.00332	360870	2381.03	269.56	−322.91
0.600	0.00350	11023	−0.00018	−36750	0.00333	360870	2441.04	255.47	−310.55
1.600	0.00238	11023	−0.00153	−322149	0.00234	360870	3526.40	−17.89	−76.72
5.000	0.00211	11023	−0.00187	−360870	0.00210	360870	3648.38	−49.56	−50.41

TABLE 1.7

P-M Interaction Values for RC Section 350×500 ($p_t = 1.5\%$, $p_c = 1.2\%$, $f_{ck} = 25$ N/mm², $f_y = 380$ N/mm²)

x_c	$\varepsilon_{c,max}$	$\sigma_{c,max}$	ε_{st}	σ_{st}	ε_{sc}	σ_{sc}	P_u	M_u+	M_u-
−1000	0.00000	0	0.01000	330435	−0.01000	−330435	−1467.63	35.88	35.88
−3.500	0.00000	0	0.01000	330435	−0.00889	−330435	−1467.63	35.88	35.88
−1.750	0.00000	0	0.01000	330435	−0.00802	−330435	−1467.63	35.88	35.88
0.000	0.00000	0	0.01000	330435	−0.00064	−134043	−1079.95	121.16	−70.74
0.024	0.00053	5038	0.01000	330435	−0.00015	−30571	−853.93	171.37	−132.17
0.051	0.00121	9325	0.01000	330435	0.00050	104812	−513.28	246.97	−222.47
0.078	0.00200	11023	0.01000	330435	0.00123	259149	−102.30	336.37	−328.63
0.094	0.00249	11023	0.01000	330435	0.00169	330435	101.10	379.27	−379.27
0.111	0.00309	11023	0.01000	330435	0.00226	330435	172.74	391.66	−391.66
0.122	0.00350	11023	0.01000	330435	0.00264	330435	217.51	398.73	−398.73
0.193	0.00350	11023	0.00504	330435	0.00296	330435	438.75	425.10	−425.10
0.264	0.00350	11023	0.00274	330435	0.00310	330435	659.98	438.42	−438.42
0.324	0.00350	11023	0.00157	330435	0.00318	330435	849.61	439.47	−439.47
0.361	0.00350	11023	0.00106	222786	0.00321	330435	1229.05	377.06	−388.75
0.419	0.00350	11023	0.00043	89499	0.00325	330435	1740.05	291.17	−317.33
0.470	0.00350	11023	0.00000	0	0.00328	330435	2120.23	223.50	−259.37
0.479	0.00350	11023	−0.00007	−13810	0.00328	330435	2182.42	211.93	−249.31
0.490	0.00350	11023	−0.00014	−29280	0.00329	330435	2253.39	198.52	−237.58
0.500	0.00350	11023	−0.00021	−44100	0.00329	330435	2322.75	185.18	−225.84
1.500	0.00233	11023	−0.00160	−330435	0.00229	330435	3378.58	−32.64	−39.12
5.000	0.00209	11023	−0.00189	−330435	0.00208	330435	3395.42	−35.64	−36.11

TABLE 1.8

P-M Interaction Values for RC Section 350×500 (p_t = 1.5%, p_c = 1.2%, f_{ck} = 25 N/mm^2, f_y = 415 N/mm^2)

x_c	$\varepsilon_{c,max}$	$\sigma_{c,max}$	ε_{st}	σ_{st}	ε_{sc}	σ_{sc}	P_u	M_u+	M_u-
−1000	0.00000	0	0.01000	360870	−0.01000	−360870	−1602.80	39.18	39.18
−3.500	0.00000	0	0.01000	360870	−0.00889	−360870	−1602.80	39.18	39.18
−1.750	0.00000	0	0.01000	360870	−0.00802	−360870	−1602.80	39.18	39.18
0.000	0.00000	0	0.01000	360870	−0.00064	−134043	−1155.05	137.69	−83.95
0.024	0.00053	5038	0.01000	360870	−0.00015	−30571	−929.03	187.89	−145.39
0.051	0.00121	9325	0.01000	360870	0.00050	104812	−588.37	263.49	−235.69
0.078	0.00200	11023	0.01000	360870	0.00123	259149	−177.40	352.89	−341.85
0.094	0.00249	11023	0.01000	360870	0.00169	354606	73.71	406.29	−405.61
0.111	0.00309	11023	0.01000	360870	0.00226	360870	157.72	421.40	−421.40
0.122	0.00350	11023	0.01000	360870	0.00264	360870	202.49	428.47	−428.47
0.190	0.00350	11023	0.00518	360870	0.00295	360870	413.88	453.94	−453.94
0.257	0.00350	11023	0.00290	360870	0.00309	360870	625.28	467.50	−467.50
0.315	0.00350	11023	0.00172	360870	0.00317	360870	806.47	469.66	−469.66
0.354	0.00350	11023	0.00115	241063	0.00320	360870	1222.94	401.19	−414.20
0.416	0.00350	11023	0.00046	95748	0.00325	360870	1774.86	308.74	−337.52
0.470	0.00350	11023	0.00000	0	0.00328	360870	2180.31	236.72	−275.90
0.479	0.00350	11023	−0.00007	−13810	0.00328	360870	2242.50	225.15	−265.83
0.490	0.00350	11023	−0.00014	−29280	0.00329	360870	2313.46	211.74	−254.10
0.500	0.00350	11023	−0.00021	−44100	0.00329	360870	2382.83	198.39	−242.36
1.500	0.00233	11023	−0.00160	−336467	0.00229	360870	3453.54	−22.69	−53.02
5.000	0.00209	11023	−0.00189	−360870	0.00208	360870	3530.59	−38.95	−39.41

TABLE 1.9

P-M Interaction Values for RC Section 350 × 600 (p_t = 1.5%, p_c = 1.2%, f_{ck} = 25 N/mm², f_y = 380 N/mm²)

x_c	$\varepsilon_{c,max}$	$\sigma_{c,max}$	ε_{st}	σ_{st}	ε_{sc}	σ_{sc}	P_u	M_u+	M_u-
−1000	0.00000	0	0.01000	330435	−0.00999	−330435	−1779.89	53.40	53.40
−3.500	0.00000	0	0.01000	330435	−0.00867	−330435	−1779.89	53.40	53.40
−1.750	0.00000	0	0.01000	330435	−0.00767	−330435	−1779.89	53.40	53.40
0.000	0.00000	0	0.01000	330435	−0.00053	−110526	−1253.43	195.54	−124.28
0.029	0.00053	5038	0.01000	330435	−0.00003	−5817	−976.35	270.89	−216.55
0.062	0.00121	9325	0.01000	330435	0.00062	131185	−559.35	383.88	−351.68
0.095	0.00200	11023	0.01000	330435	0.00137	287368	−56.51	517.33	−510.37
0.113	0.00249	11023	0.01000	330435	0.00183	330435	122.61	562.49	−562.49
0.135	0.00309	11023	0.01000	330435	0.00240	330435	209.49	580.44	−580.44
0.148	0.00350	11023	0.01000	330435	0.00279	330435	263.79	590.66	−590.66
0.234	0.00350	11023	0.00504	330435	0.00305	330435	532.10	628.58	−628.58
0.320	0.00350	11023	0.00274	330435	0.00317	330435	800.40	647.33	−647.33
0.393	0.00350	11023	0.00157	330435	0.00323	330435	1030.38	648.13	−648.13
0.437	0.00350	11023	0.00106	222786	0.00326	330435	1490.55	554.87	−572.27
0.508	0.00350	11023	0.00043	89499	0.00329	330435	2110.27	426.57	−465.50
0.570	0.00350	11023	0.00000	0	0.00332	330435	2571.35	325.56	−378.96
0.579	0.00350	11023	−0.00005	−11425	0.00332	330435	2633.64	311.33	−366.57
0.590	0.00350	11023	−0.00012	−24313	0.00332	330435	2705.01	294.82	−352.14
0.600	0.00350	11023	−0.00018	−36750	0.00333	330435	2775.02	278.38	−337.72
1.600	0.00238	11023	−0.00153	−322149	0.00234	330435	4041.27	−40.54	−64.91
5.000	0.00211	11023	−0.00187	−330435	0.00210	330435	4092.50	−52.90	−53.89

TABLE 1.10

P-M Interaction Values for RC Section 350×600 ($p_t = 1.5\%$, $p_c = 1.2\%$, $f_{ck} = 25$ N/mm^2, $f_y = 415$ N/mm^2)

x_c	$\varepsilon_{c,max}$	$\sigma_{c,max}$	ε_{st}	σ_{st}	ε_{sc}	σ_{sc}	P_u	M_u+	M_u-
−1.000	0.00000	0	0.01000	360870	−0.00999	−360870	−1943.82	58.31	58.31
−3.500	0.00000	0	0.01000	360870	−0.00867	−360870	−1943.82	58.31	58.31
−1.750	0.00000	0	0.01000	360870	−0.00767	−360870	−1943.82	58.31	58.31
0.000	0.00000	0	0.01000	360870	−0.00053	−110526	−1344.50	220.13	−143.96
0.029	0.00053	5038	0.01000	360870	−0.00003	−5817	−1067.43	295.48	−236.22
0.062	0.00121	9325	0.01000	360870	0.00062	131185	−650.42	408.47	−371.35
0.095	0.00200	11023	0.01000	360870	0.00137	287368	−147.59	541.92	−530.05
0.113	0.00249	11023	0.01000	360870	0.00183	360870	104.39	606.75	−606.75
0.135	0.00309	11023	0.01000	360870	0.00240	360870	191.27	624.70	−624.70
0.148	0.00350	11023	0.01000	360870	0.00279	360870	245.57	634.93	−634.93
0.230	0.00350	11023	0.00518	360870	0.00304	360870	501.94	671.57	−671.57
0.312	0.00350	11023	0.00290	360870	0.00316	360870	758.31	690.70	−690.70
0.382	0.00350	11023	0.00172	360870	0.00323	360870	978.06	693.16	−693.16
0.429	0.00350	11023	0.00115	241063	0.00326	360870	1483.14	590.86	−610.22
0.504	0.00350	11023	0.00046	95748	0.00329	360870	2152.50	452.74	−495.58
0.570	0.00350	11023	0.00000	0	0.00332	360870	2644.21	345.24	−403.55
0.579	0.00350	11023	−0.00005	−11425	0.00332	360870	2706.51	331.00	−391.17
0.590	0.00350	11023	−0.00012	−24313	0.00332	360870	2777.87	314.49	−376.73
0.600	0.00350	11023	−0.00018	−36750	0.00333	360870	2847.88	298.05	−362.31
1.600	0.00238	11023	−0.00153	−322149	0.00234	360870	4114.13	−20.87	−89.50
5.000	0.00211	11023	−0.00187	−360870	0.00210	360870	4256.44	−57.82	−58.81

TABLE 1.11

P-M Interaction Values for RC Section 350×700 ($p_t = 1.5\%$, $p_c = 1.2\%$, $f_{ck} = 25$ N/mm^2, $f_y = 380$ N/mm^2)

x_c	$\varepsilon_{c,max}$	$\sigma_{c,max}$	ε_{st}	σ_{st}	ε_{sc}	σ_{sc}	P_u	M_u+	M_u-
−1000	0.00000	0	0.01000	330435	−0.00999	−330435	−2092.15	74.39	74.39
−3.500	0.00000	0	0.01000	330435	−0.00847	−330435	−2092.15	74.39	74.39
−1.750	0.00000	0	0.01000	330435	−0.00736	−330435	−2092.15	74.39	74.39
0.000	0.00000	0	0.01000	330435	−0.00045	−94030	−1426.90	287.27	−191.71
0.034	0.00053	5038	0.01000	330435	0.00005	11548	−1098.78	392.84	−321.05
0.073	0.00121	9325	0.01000	330435	0.00071	149686	−605.42	550.72	−510.03
0.112	0.00200	11023	0.01000	330435	0.00146	307164	−10.72	737.03	−731.79
0.133	0.00249	11023	0.01000	330435	0.00193	330435	144.12	781.68	−781.68
0.158	0.00309	11023	0.01000	330435	0.00250	330435	246.24	806.21	−806.21
0.174	0.00350	11023	0.01000	330435	0.00290	330435	310.07	820.17	−820.17
0.275	0.00350	11023	0.00504	330435	0.00312	330435	625.45	871.73	−871.73
0.376	0.00350	11023	0.00274	330435	0.00322	330435	940.83	896.80	−896.80
0.462	0.00350	11023	0.00157	330435	0.00327	330435	1211.15	897.20	−897.20
0.514	0.00350	11023	0.00106	222786	0.00330	330435	1752.05	766.92	−791.16
0.597	0.00350	11023	0.00043	89499	0.00332	330435	2480.50	587.74	−641.98
0.670	0.00350	11023	0.00000	0	0.00334	330435	3022.46	446.76	−521.15
0.679	0.00350	11023	−0.00005	−9742	0.00335	330435	3084.84	429.86	−506.44
0.690	0.00350	11023	−0.00010	−20787	0.00335	330435	3156.48	410.24	−489.30
0.700	0.00350	11023	−0.00015	−31500	0.00335	330435	3226.96	390.70	−472.18
1.700	0.00243	11023	−0.00147	−309000	0.00239	330435	4675.50	−39.76	−104.19
5.000	0.00213	11023	−0.00184	−330435	0.00211	330435	4789.16	−73.46	−75.32

TABLE 1.12

P-M Interaction Values for RC Section 350 × 700 ($p_t = 1.5\%$, $p_c = 1.2\%$, $f_{ck} = 25$ N/mm², $f_y = 415$ N/mm²)

x_c	$\varepsilon_{c,max}$	$\sigma_{c,max}$	ε_{st}	σ_{st}	ε_{sc}	σ_{sc}	P_u	M_u+	M_u-
-1000	0.00000	0	0.01000	360870	-0.00999	-360870	-2284.85	81.24	81.24
-3.500	0.00000	0	0.01000	360870	-0.00847	-360870	-2284.85	81.24	81.24
-1.750	0.00000	0	0.01000	360870	-0.00736	-360870	-2284.85	81.24	81.24
0.000	0.00000	0	0.01000	360870	-0.00045	-94030	-1533.96	321.52	-219.12
0.034	0.00053	5038	0.01000	360870	0.00005	11548	-1205.83	427.10	-348.46
0.073	0.00121	9325	0.01000	360870	0.00071	149686	-712.47	584.98	-537.44
0.112	0.00200	11023	0.01000	360870	0.00146	307164	-117.78	771.29	-759.20
0.133	0.00249	11023	0.01000	360870	0.00193	360870	122.71	843.34	-843.34
0.158	0.00309	11023	0.01000	360870	0.00250	360870	224.83	867.88	-867.88
0.174	0.00350	11023	0.01000	360870	0.00290	360870	288.66	881.84	-881.84
0.270	0.00350	11023	0.00518	360870	0.00311	360870	590.00	931.67	-931.67
0.367	0.00350	11023	0.00290	360870	0.00321	360870	891.35	957.31	-957.31
0.449	0.00350	11023	0.00172	360870	0.00327	360870	1149.65	960.03	-960.03
0.505	0.00350	11023	0.00115	241063	0.00329	360870	1743.34	817.12	-844.09
0.593	0.00350	11023	0.00046	95748	0.00332	360870	2530.13	624.21	-683.90
0.670	0.00350	11023	0.00000	0	0.00334	360870	3108.10	474.16	-555.40
0.679	0.00350	11023	-0.00005	-9742	0.00335	360870	3170.48	457.26	-540.70
0.690	0.00350	11023	-0.00010	-20787	0.00335	360870	3242.12	437.64	-523.56
0.700	0.00350	11023	-0.00015	-31500	0.00335	360870	3312.60	418.11	-506.44
1.700	0.00243	11023	-0.00147	-309000	0.00239	360870	4761.14	-12.36	-138.44
5.000	0.00213	11023	-0.00184	-360870	0.00211	360870	4981.86	-80.31	-82.17

TABLE 1.13
Summary of Expressions for P-M Interaction Behavior

(1)	$[-\infty, 0]$	0	$\dfrac{\varepsilon_{su} x_c}{D - x_c - d}$	ε_{su}	$\varepsilon_{su}\left(\dfrac{x_c - d}{D - x_c - d}\right)$	$P_{u,st}$	$M_{u,st}$
(2a)	$[0, x'_{c,lim}]$	0	$\dfrac{\varepsilon_{su} x_c}{D - x_c - d}$	ε_{su}	$\varepsilon_{su}\left(\dfrac{x_c - d}{D - x_c - d}\right)$	$P_{u,st} + P_{u1,con}$ $(q=0)$	$M_{u,st} +$ $M_{u1,con}(q=0)$
(2b)	$[x'_{c,lim}, x''_{c,lim}]$	$x_c - \dfrac{\varepsilon_{c0}}{\varepsilon_{su}}(D - x_c - d)$	$\dfrac{\varepsilon_{su} x_c}{D - x_c - d}$	ε_{su}	$\varepsilon_{su}\left(\dfrac{x_c - d}{D - x_c - d}\right)$	$P_{u,st} + P_{u1,con}$	$M_{u,st} + M_{u1,con}$
(3)	$[x'_{c,lim}, x''_{c,lim}]$	$\dfrac{\varepsilon_{cu} - \varepsilon_{c0}}{\varepsilon_{cu}} x_c$	ε_{su}	$\dfrac{\varepsilon_{cu}(D - x_c - d)}{x_c}$	$\dfrac{\varepsilon_{cu}(x_c - d)}{x_c}$	$P_{u,st} + P_{u1,con}$	$M_{u,st} + M_{u1,con}$
(4)	$[x''_{c,lim}, (D-d)]$	$\dfrac{\varepsilon_{cu} - \varepsilon_{c0}}{\varepsilon_{cu}} x_c$	ε_{su}	$\dfrac{\varepsilon_{cu}(D - x_c - d)}{x_c}$	$\dfrac{\varepsilon_{cu}(x_c - d)}{x_c}$	$P_{u,st} + P_{u1,con}$	$M_{u,st} + M_{u1,con}$
(5)	$[(D-d), D]$	$\dfrac{\varepsilon_{cu} - \varepsilon_{c0}}{\varepsilon_{cu}} x_c$	ε_{su}	$\dfrac{\varepsilon_{cu}(D - x_c - d)}{x_c}$	$\dfrac{\varepsilon_{cu}(x_c - d)}{x_c}$	$P_{u,st} + P_{u1,con}$	$M_{u,st} + M_{u1,con}$
(6)	$[D, +\infty[$	$\dfrac{\varepsilon_{cu} - \varepsilon_{c0}}{\varepsilon_{cu}} D$	$\dfrac{\varepsilon_{cu}\varepsilon_{c0} x_c}{\varepsilon_{cu} x_c - D(\varepsilon_{cu} - \varepsilon_{c0})}$	$\dfrac{\varepsilon_{cu}\varepsilon_{c0}(D - x_c - d)}{\varepsilon_{cu} x_c - D(\varepsilon_{cu} - \varepsilon_{c0})}$	$\dfrac{\varepsilon_{cu}\varepsilon_{c0}(x_c - d)}{\varepsilon_{cu} x_c - D(\varepsilon_{cu} - \varepsilon_{c0})}$	$P_{u,st} + P_{u1,con} +$ $P_{u2,con}$	$M_{u,st} + M_{u1,con}$ $+ M_{u2,con}$

(Continued)

TABLE 1.13
Summary of Expressions for P-M Interaction Behavior (Continued)

where $\varepsilon_{c.max}(x_c) = \varepsilon_c(x_c, y = 0)$

$$P_{u,st} = A_{sc}\sigma_{sc} - A_{st}\sigma_{st} = b(D-d)(p_c\sigma_{sc} - p_t\sigma_{st})$$

$$M_{u,st} = (A_{sc}\sigma_{sc} - A_{st}\sigma_{st})\left(\frac{D}{2} - d\right) = b(D-d)(p_c\sigma_{sc} - p_t\sigma_{st})\left(\frac{D}{2} - d\right)$$

$$P_{u1,con} = bq\sigma_{c0} + \frac{b\sigma_{c0}\varepsilon_{c.max}(q-x_c)^2}{3x_c^2\varepsilon_{c0}^2}[3\varepsilon_{c0}x_c + \varepsilon_{c.max}(q-x_c)]$$

$$M_{u1,con} = \frac{bq\sigma_{c0}}{2}(D-q) - \frac{b\sigma_{c0}\varepsilon_{c.max}(q-x_c)^2}{12x_c^2\varepsilon_{c0}^2}[2\varepsilon_{c0}x_c(4q+2x_c-3D)+\varepsilon_{c.max}(q-x_c)(x_c+3q-2D)]$$

$$P_{u2,con} = -\frac{b\sigma_{c0}\varepsilon_{c.max}[3\varepsilon_{c0}x_c + \varepsilon_{c.max}(D-x_c)](D-x_c)^2}{3x_c^2\varepsilon_{c0}^2}$$

$$M_{u2,con} = \frac{b\sigma_{c0}\varepsilon_{c.max}(D-x_c)^2[2\varepsilon_{c0}x_c(D+2x_c)+\varepsilon_{c.max}(D^2-x_c^2)]}{12x_c^2\varepsilon_{c0}^2}$$

1.6 CONCLUSIONS

In this chapter, a detailed methodology for estimating the P-M yield interaction is presented while identifying the subdomains of the P-M boundary. RC beams of different cross-sections and percentage of reinforcement in tension and compression are analyzed, and their subdomains in P-M interaction are identified. The proposed expressions for P-M and M-ϕ are carefully examined for their close agreement with selected examples of RC beams. Though some of the observations are already reported in the literature, the study quantifies the value through illustrated examples relevant to the Eurocode currently prevalent. The expressions presented in a closed form will be very useful to the engineering community to perform nonlinear analyses like pushover.

1.7 NUMERICAL PROCEDURE IN SPREADSHEET FORMAT

A compact disc with relevant content can be downloaded free from http://www.crcpress.com/e_products/downloads/download.asp?cat_no=K10453. The user can change the common design parameters (shown in color in the spreadsheet), namely, (1) diameter and number of bars of tensile and compression reinforcement, (2) cross-section dimensions, and (3) material properties like f_{ck} and f_y. All relative coefficients required for each domain are computed automatically, and one can find the required P-M curve plotted.

2 Moment-Curvature Relationship for RC Sections

2.1 SUMMARY

The correct estimate of curvature ductility of reinforced concrete members has always been an attractive subject of study because it engenders a reliable estimate of the capacity of buildings under seismic loads. The majority of building stock needs structural assessment to certify its safety under revised seismic loads by new codes. Structural assessment of existing buildings, by employing nonlinear analyses tools like pushover, needs an accurate input of moment-curvature relationship for reliable results. In this chapter, analytical predictions of curvature ductility of reinforced concrete sections are presented. Relationships, in explicit form, to estimate the moment-curvature response are proposed, leading to closed form solutions after their verification with those obtained from numerical procedures. The purpose is to estimate curvature ductility under service loads in a simpler closed form. The influence of longitudinal tensile and compression reinforcements on curvature ductility is also examined and discussed. The spreadsheet program used to estimate the moment-curvature relationship, after simplifying the complexities involved in such estimates, predicts in good agreement with the proposed analytical expressions. In lieu of tedious hand calculations and approximations required in conventional iterative design procedures, the proposed estimate of curvature ductility provides a ready solution for a potentially safe design.

2.2 INTRODUCTION

Earthquake-resistant design of RC-framed structures is essentially focused on the displacement ductility of buildings instead of on material ductility of reinforcing bars. Critical points of interest are the strain levels in concrete and steel, indicating whether the failure is tensile or compressive at the instant of reaching plastic hinge formation (Pisanty and Regan 1998). Estimate of ductility demand is of particular interest to structural designers for ensuring effective redistribution of moments in ultraelastic response, allowing for the development of energy dissipative zones until collapse (Pisanty and Regan 1993). In seismic areas, ductility of the structure is an important design parameter since modern seismic design philosophy is based on energy absorption and dissipation by postelastic deformation for the survival of the structure during major earthquakes (Park and Kim 2003). Studies conducted on existing buildings showed they were structurally unfit to support seismic loads demanded by the revised international codes (see, for example, Chandrasekaran and

Roy 2006; Chao et al. 2006). The deformation demand predictions by improved Demand Capacity Method are sensitive to ductility, since higher ductility results in conservative predictions (Sinan and Asli 2007). The estimate of moment-curvature relationship of RC sections has been a point of research interest for many years (see, for example, Pfrang, Siess, and Sozen 1964; Carreira and Chu 1986; Mo 1992); historically, moment-curvature relationships with softening branch were first introduced by Wood (1968). Load-deformation characteristics of reinforced concrete members, bending in particular, are mainly dependent on moment-curvature characteristics of the sections since most of these deformations arise from strains associated with flexure (Park and Paulay 1975). Studies also show that in well-designed and detailed RC structures, the gap between the actual and design lateral forces narrows down by ensuring ductility (see, for example, Wood 1968; Pankaj and Manish 2006). With regard to RC building frames with sidesway, their response assessment is complicated because of second-order deformations and because considerable redistribution of moments may occur as a result of plastic behavior of sections as well (Nunziante and Ocone 1988). Plastic curvature is therefore a complex issue mainly because of interaction of various parameters, namely, (1) the response of constitutive material, (2) member geometry, and (3) loading conditions. Observations made on plastic softening beams (Challamel and Hjiaj 2005) showed that the correct estimate of yield moment, a nonlocal material parameter, is important to ensure proper continuity between elastic and plastic regions during the loading process. Experimental evidence of the moment-curvature relationship of RC sections already faced limited loading cases and support conditions (Ko, Kim, and Kim 2001). Mo (1992) performed elastic-plastic buckling analysis by employing a finite element procedure to reproduce moment-curvature relationship with the softening branch, and Jirasek and Bazant (2002) used an alternate, simplified model where the complex nonlinear geometric effects are embedded in the developed model of material behavior. Experimental investigations also impose limitations in estimating the plastic rotation capacity. As already seen (Lopes and Bernardo 2003), the experimental results for rotation-deflection behavior showed good agreement with the analysis in the elastic regime, but for the phase of yielding of reinforcement steel, the theoretical results did not agree with the experimental inferences.

Studies reported above show that no simplified procedure exists to estimate curvature ductility of RC sections. The response of RC building frames under ground shaking generally results in nonlinear behavior, which is complex to model. Further, increased use of a displacement-based design approach leads to nonlinear static procedures for estimating seismic demands (FEMA 450, 2004; FEMA 440, 2005) for which such an estimate of moment-curvature relationship is essential. Therefore, in this study, a simplified numerical procedure for moment-curvature relationship of RC sections is attempted. The computations are based on their nonlinear characteristics in full depth of the cross-section, for different ratios of longitudinal tensile and compression reinforcements. They account for the variation on depth of neutral axis passing through different domains, classified on the basis of strain levels reached in the constitutive materials, namely, concrete and steel. Obtained results, by employing the numerical procedure on example RC sections, are verified with expressions derived from detailed analytical modeling.

2.3 MATHEMATICAL DEVELOPMENT

Significant nonlinearity exhibited by concrete, under multiaxial stress state, can be successively represented by nonlinear characteristics of constitutive models capable of interpreting inelastic deformations (Chen 1994a, 1994b). Elastic stress-strain relationship of constitutive materials, as prescribed by the code currently in prevalence (D.M. 9 gennaio 1996; Eurocode UNI ENV 1991-1, 1991-2; Ordinanza 2003, 2005; D.M. 2005) are used in this study, as already presented in Chapter 1. The fundamental Bernoulli's hypothesis of linear strain over the cross-section, both for elastic and elastic-plastic responses of the beam under bending moment combined with axial force, will be assumed. The interaction behavior becomes critical when one of the following conditions applies: (1) strain in reinforcing steel in tension reaches ultimate limit; (2) strain in concrete in extreme compression fiber reaches ultimate limit; or (3) maximum strain in concrete in compression reaches elastic limit under only axial compression.

2.4 MOMENT-CURVATURE IN ELASTIC RANGE

It is well known that the bending curvature is the derivative of bending rotation, varying along the member length and at any cross-section, and is given by slope of the strain profile. It depends on the fluctuations of neutral axis depth and continuously varying strains. Moment-curvature relationship, in elastic range, depends on both the magnitude and nature of the axial force. Figure 2.1 shows the curvature

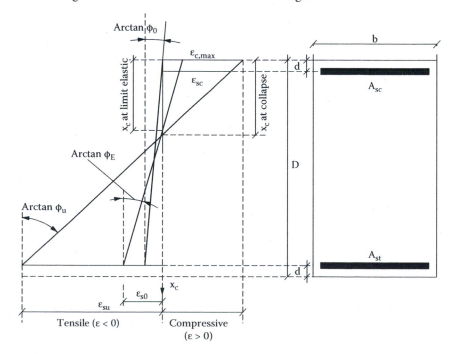

FIGURE 2.1 Curvature profile for strain variation in concrete and steel.

profile for strain variation in concrete and steel. Magnitude of axial force is assumed to vary in the range of

$$-(A_{sc} + A_{st}) \sigma_{s0} < P < \{bD\sigma_{c0} + (A_{sc} + A_{st}) \sigma_{s0}\} \qquad (2.1a)$$

The nature of axial force will vary as (1) tensile axial force (considered as negative in this study); (2) zero axial force; as well as (3) compressive axial force (considered positive). Stress and strain in concrete and steel, in elastic range are given by

$$\varepsilon_c = \varphi_e(x_c - y); \quad \varepsilon_{sc} = \varphi_e(x_c - d); \quad \varepsilon_{st} = \varphi_e(D - x_c - d);$$

$$\sigma_c = \frac{(x_c - y)[2\varepsilon_{c0} - (x_c - y)\varphi_e]\sigma_{c0}\varphi_e}{\varepsilon_{c0}^2}; \quad \sigma_{sc} = E_s\varphi_e(x_c - d); \qquad (2.1b)$$

$$\sigma_{st} = E_s\varphi_e(D - x_c - d);$$

2.4.1 TENSILE AXIAL FORCE

Tensile axial force results in reduced curvature. Under the action of axial force, the equilibrium equation for axial force is written along the length axis of the beam while moment is evaluated about the CG of the cross-section. Expressions for axial force and bending moment, in explicit form, are given by

$$P_e = -\sigma_{st}A_{st} + \sigma_{sc}A_{sc} = b(d - D)[d(p_c - p_t) + Dp_t - (p_c + p_t)x_c]E_s\phi \qquad (2.2)$$

$$M_e = (\sigma_{st}A_{st} + \sigma_{sc}A_{sc})\left(\frac{D}{2} - d\right)$$

$$= \frac{1}{2}b(D - 2d)(D - d)[(p_c - p_t)x_c + Dp_t - d(p_c + p_t)]E_s\phi \qquad (2.3)$$

Percentages of steel, in tension and compression zones, are given by

$$A_{st} = p_t b(D - d); \quad A_{sc} = p_c b(D - d) \qquad (2.4)$$

By solving Equation 2.2 with respect to x_c, we obtain the following relationship:

$$x_c = \frac{P_e + b(d - D)[d(p_c - p_t) + Dp_t]E_s\phi}{b(d - D)(p_c + p_t)E_s\phi} \qquad (2.5)$$

By substituting Equation 2.5 in Equation 2.3, moment-curvature relationship is obtained as

$$M_c = \frac{D-2d}{2(p_c + p_t)}\left[P_e(p_c - p_t) + 2b(D^2 + d^2 - 3dD)E_s p_c p_t \phi\right] \quad \forall \; \phi \in [0, \phi_0] \quad (2.6)$$

where ϕ_0 is the limit curvature for $x_c = 0$; by imposing this condition in Equation 2.5, we get

$$\phi_0 = \frac{P_e}{b(d-D)E_s[Dp_t + d(p_c - p_t)]} \quad (2.7)$$

As curvature is influenced by the percentage of tension reinforcement, by imposing the conditions $x_c = 0$ and $\phi = \varepsilon_{s0}/(D-d)$ in Equation 2.2 and solving with respect to p_t, for a specified range of $\phi_u^{(b)} = -Q_1 + \sqrt{Q_1^2 - 4Q_0 Q_2}/2Q_2$, Equation 2.6 is defined in the total range $[0, \phi_E]$, where ϕ_E is the limit elastic curvature and is derived in the following section. For further increase in curvature more than ϕ_0, concrete also contributes to the resulting compression, and expressions for axial force and bending moment take the following form:

$$P_e = \int_0^{x_c} b\sigma_c[\varepsilon_c(y)]dy - \sigma_{st}A_{st} + \sigma_{sc}A_{sc} = A_0(\phi_e) + A_1(\phi_e)x_c + A_2(\phi_e)x_c^2 + A_3(\phi_e)x_c^3$$

$$(2.8)$$

$$M_c = \int_0^{x_c} b\sigma_c[\varepsilon_c(y)]\left(\frac{D}{2} - y\right)dy + (\sigma_{st}A_{st} + \sigma_{sc}A_{sc})\left(\frac{D}{2} - d\right)$$

$$M_c = B_0(\phi_e) + B_1(\phi_e)x_c + B_2(\phi_e)x_c^2 + B_3(\phi_e)x_c^3 + B_4(\phi_e)x_c^4 \quad (2.9)$$

where the coefficients A_i (for $i = 0$ to 3) and B_i (for $i = 0$ to 4), as a function of curvature, are given by

$$A_0(\phi_e) = b(d-D)[Dp_t + d(p_c - p_t)]E_s\phi_e; \quad A_1(\phi_e) = b(D-d)(p_c + p_t)E_s\phi_e;$$

$$A_2(\phi_e) = \frac{b\sigma_{c0}\phi_e}{\varepsilon_{c0}}; \quad A_3(\phi_e) = -\frac{b\sigma_{c0}\phi_e^2}{3\varepsilon_{c0}^2};$$

$$B_0(\phi_e) = \frac{1}{2}b(2d^2 - 3dD + D^2)[Dp_t - d(p_c + p_t)]E_s\phi_e; \quad (2.10)$$

$$B_1(\phi_e) = \frac{1}{2}b(2d^2 - 3dD + D^2)(p_c - p_t)E_s\phi_e; \quad B_2(\phi_e) = \frac{bD\sigma_{c0}\phi_e}{2\varepsilon_{c0}};$$

$$B_3(\phi_e) = -\frac{b\sigma_{c0}\phi_e(2\varepsilon_{c0} + D\phi_e)}{6\varepsilon_{c0}^2}; \quad B_4(\phi_e) = \frac{b\sigma_{c0}\phi_e^2}{12\varepsilon_{c0}^2};$$

By solving Equation 2.8 with respect to variable x_c, three roots of the variable are obtained as

$$x_{c1}(P_e,\phi_e) = \frac{1}{6A_3(\phi_e)}$$

$$\times\left[-2A_2(\phi_e) + \frac{2.5198\left(A_2^2(\phi_e) - 3A_1(\phi_e)A_3(\phi_e)\right)}{C_1(\phi_e,P_e)} + 1.5874\,C_1(\phi_e,P_e)\right]$$

$$x_{c2}(P_e,\phi_e) = \frac{1}{12A_3(\phi_e)}\left[\begin{array}{l} -4A_2(\phi_e) - \dfrac{(2.5198 + 4.3645\,i)\left(A_2^2(\phi_e) - 3A_1(\phi_e)A_3(\phi_e)\right)}{C_1(\phi_e,P_e)} \\[2mm] -(1.5874 - 2.7495\,i)C_1(\phi_e,P_e) \end{array}\right]$$

$$x_{c3}(P_e,\phi_e) = \frac{1}{12A_3(\phi_e)}\left[\begin{array}{l} -4A_2(\phi_e) - \dfrac{(2.5198 - 4.3645\,i)\left(A_2^2(\phi_e) - 3A_1(\phi_e)A_3(\phi_e)\right)}{C_1(\phi_e,P_e)} \\[2mm] -(1.5874 + 2.7495\,i)C_1(\phi_e,P_e) \end{array}\right]$$

$$(2.11)$$

where,

$$C_1(\phi_e,P_e) = \left[\begin{array}{l} +\sqrt{-4\left(A_2^2 - 3A_1A_3\right)^3 + \left(2A_2^3 - 9A_1A_2A_3 + 27A_3^2(A_0 - P_e)\right)^2} \\[2mm] -2A_2^3 + 9A_1A_2A_3 - 27A_3^2(A_0 - P_e) \end{array}\right]^{1/3} \quad (2.12)$$

Out of the above, only one root, namely x_{c3}, is in close agreement with the obtained numerical solution. By substituting the root x_{c3} in Equation 2.11, the moment-curvature relationship in elastic range is obtained as

$$M_e = B_0(\phi_e) + B_1(\phi_e)x_{c3}(\phi_e,P_e) + B_2(\phi_e)x_{c3}^2(\phi_e,P_e)$$

$$+ B_3(\phi_e)x_{c3}^3(\phi_e,P_e) + B_4(\phi_e)x_{c3}^4(\phi_e,P_e) \qquad \forall\phi \in [\phi_0,\phi_E]$$

$$(2.13)$$

2.4.2 NO AXIAL FORCE

The moment-curvature relationship is given by Equation 2.13 for the complete range of $[0,\phi_E]$.

2.4.3 COMPRESSIVE AXIAL FORCE

Expressions for axial force and bending moment are given by

$$P_e = \int_0^D b\sigma_c[\varepsilon_c(y)]\,dy - \sigma_{st}A_{st} + \sigma_{sc}A_{sc} = E_0 + E_1 x_c + E_2 x_c^2 \qquad (2.14)$$

$$M_e = \int_0^D b\sigma_c[\varepsilon_c(y)]\left(\frac{D}{2}-y\right)dy + (\sigma_{st}A_{st}+\sigma_{sc}A_{sc})\left(\frac{D}{2}-d\right) = F_0 + F_1 x_c \qquad (2.15)$$

where the coefficients $E_{i=0,1,2}$ and $F_{i=0,1}$ are given by

$$E_0 = \frac{1}{3}b\phi\left[3d(d-D)E_sp_c - 3(d-D)^2E_sp_t - \frac{D^2\sigma_{c0}(3\varepsilon_{c0}+D\phi)}{\varepsilon_{c0}^2}\right],$$

$$E_1 = \frac{b\phi\left[-dE_s(p_c+p_t)\varepsilon_{c0}^2 + D(E_s(p_c+p_t)\varepsilon_{c0}^2 + \sigma_{c0}(2\varepsilon_{c0}+D\phi))\right]}{\varepsilon_{c0}^2},$$

$$E_2 = -\frac{bD\sigma_{c0}\phi^2}{\varepsilon_{c0}^2},$$

$$F_0 = -\frac{b\phi}{12}\left[6d(D-2d)(D-d)E_sp_c + 6(d-D)^2(2d-D)E_sp_t - \frac{D^3\sigma_{c0}(2\varepsilon_{c0}+D\phi)}{\varepsilon_{c0}^2}\right],$$

$$F_1 = \frac{b\phi\left[3(D^2+2d^2-3dD)(p_c-p_t)E_s\varepsilon_{c0}^2 - D^3\sigma_{c0}\phi\right]}{6\varepsilon_{c0}^2}$$

$$(2.16)$$

By solving Equation 2.14, the position of the neutral axis is determined as

$$x_c = \frac{-E_1 + \sqrt{E_1^2 - 4E_2(E_0-P_e)}}{2E_0} \qquad (2.17)$$

By substituting Equation 2.17 in Equation 2.15, we get

$$M_e = F_0(\phi,P_e) + F_1(\phi,P_e)x_c \quad \forall \phi \in [0,\phi_0] \qquad (2.18)$$

where

$$\phi_0 = \frac{3b\varepsilon_{c0}\left[(D-d)E_s(Dp_c+d(p_t-p_c)) + D^2\sigma_{c0}\right]}{2bD^3\sigma_{c0}}$$

$$- \frac{\varepsilon_{c0}\sqrt{3b}\sqrt{3b\left[(D-d)E_s\varepsilon_{c0}\left((Dp_c+d(p_t-p_c)) + D^2\sigma_{c0}\right)^2\right] - 4P_eD^3\sigma_{c0}}}{2bD^3\sigma_{c0}}$$

$$(2.19)$$

By imposing the condition ($x_c = D$) in Equation 2.17, limit curvature ϕ_0 is determined as given above. Further increase in curvature changes the equilibrium conditions due to contributions from the resultant compressive force by concrete. For curvature more than ϕ_0, moment-curvature relationship is discussed in the next section.

2.5 ELASTIC LIMIT BENDING MOMENT AND CURVATURE

The limit elastic curvature, depending on the magnitude of axial force and percentage of reinforcing steel in tension and compression, results in four possible cases, namely, (1) strain in tension steel reaches yield limit and stress in concrete vanishes; (2) strain in tension steel reaches yield limit but stress in concrete is present; (3) strain in compression steel reaches elastic limit; and (4) strain in extreme compression fiber in concrete reaches elastic limit value.

2.5.1 CASE 1: STRAIN IN TENSION STEEL REACHES YIELD LIMIT AND STRESS IN CONCRETE VANISHES

This case is verified when $p_t < P_e + bdE_s p_c \varepsilon_{s0}/b(d-D)\sigma_{s0}$. By imposing $\sigma_{st} = \sigma_{s0}$ and recalling Equation 2.2, the depth of the neutral axis can be obtained as given below:

$$x_c^{(i)} = D - d - \frac{\varepsilon_{s0}}{\phi_E} \quad \forall x_c < 0 \tag{2.20}$$

By substituting Equation 2.20 in Equation 2.2, elastic limit curvature can be determined as

$$\phi_E = \frac{P_E + b(D-d)(p_c + p_t)\sigma_{s0}}{bE_s p_c(D^2 + 2d^2 - 3dD)} \tag{2.21}$$

By substituting Equation 2.21 in Equation 2.3, elastic limit moment is obtained as

$$M_E^{(i)} = \frac{D - 2d}{2}[P_E + 2b(D-d)p_t\sigma_{s0}] \tag{2.22}$$

2.5.2 CASE 2: STRAIN IN TENSION STEEL REACHES YIELD LIMIT AND STRESS IN CONCRETE DOES NOT EQUAL ZERO

Depth of neutral axis is given by

$$x_c^{(ii)} = D - d - \frac{\varepsilon_{s0}}{\phi_E} \quad \forall x_c \in [0, D-d] \tag{2.23}$$

By substituting Equation 2.23 in Equation 2.8, the expression for limit elastic curvature can be obtained as

$$L_0 + L_1 \phi_E + L_2 \phi_E^2 + L_3 \phi_E^3 = 0 \tag{2.24}$$

where the coefficients $L_{i=0,1,2,3}$ are given by

$$L_0 = \frac{b\varepsilon_{s0}^2(3\varepsilon_{c0}+\varepsilon_{s0})\sigma_{c0}}{3\varepsilon_{c0}^2},$$

$$L_1 = -\frac{P_E\varepsilon_{c0}^2 + b(D-d)\varepsilon_{s0}\left[(2\varepsilon_{c0}+\varepsilon_{s0})\sigma_{c0}+E_s(p_c+p_t)\varepsilon_{c0}^2\right]}{\varepsilon_{c0}^2},$$

(2.25)

$$L_2 = \frac{b(D-d)\left[(2d-D)E_sp_c\varepsilon_{c0}^2 + (d-D)(\varepsilon_{c0}+\varepsilon_{s0})\sigma_{c0}\right]}{\varepsilon_{c0}^2},$$

$$L_3 = \frac{b(d-D)^3\sigma_{c0}}{3\varepsilon_{c0}^2}$$

By solving Equation 2.24, which is of a third-degree polynomial, only one real root (third root) gives the limit elastic curvature:

$$\phi_E^{(ii)} = \frac{1}{12L_3}\left[-4L_2 - \frac{(2.5198-4.3645i)(L_2^2-3L_1L_3)}{\lambda} - (1.5874+2.7495i)\lambda\right]$$

(2.26)

where

$$\lambda = \left[-2L_2^3 + 9L_1L_2L_3 - 27L_3^2L_0 + \sqrt{-4(L_2^2-3L_1L_3)^3 + (2L_2^3-9L_1L_2L_3+27L_3^2L_0)^2}\right]^{1/3}$$

(2.27)

By substituting Equation 2.26 in Equation 2.9, limit elastic bending moment is obtained as:

$$M_E^{(ii)} = \frac{b}{2\varepsilon_{c0}^2}\left[\frac{M_1^{(ii)}}{\phi_E^{(ii)2}} + \frac{M_2^{(ii)}}{\phi_E^{(ii)}} + M_3^{(ii)} + M_4^{(ii)}\phi_E^{(ii)} + M_5^{(ii)}\phi_E^{(ii)2}\right]$$

(2.28)

where the superscript $^{(ii)}$ represents the second case. Constants of the above equation are given by

$$M_1^{(ii)} = \frac{\varepsilon_{s0}^3(4\varepsilon_{c0}+\varepsilon_{s0})\sigma_{c0}}{6}, \quad M_2^{(ii)} = -\frac{(D-2d)\varepsilon_{s0}^2(3\varepsilon_{c0}+\varepsilon_{s0})\sigma_{c0}}{3},$$

$$M_3^{(ii)} = (D-d)\varepsilon_{s0}\left[(D-2d)E_s(p_t-p_c)\varepsilon_{c0}^2 - d(2\varepsilon_{c0}+\varepsilon_{s0})\sigma_{c0}\right],$$

$$M_4^{(ii)} = \frac{(D-d)\left[3(D-2d)^2E_sp_c\varepsilon_{c0}^2 + (D-d)(2d+D)(2\varepsilon_{c0}+\varepsilon_{s0})\sigma_{c0}\right]}{3},$$

(2.29)

$$M_5^{(ii)} = \frac{(d-D)^3(D+d)\sigma_{c0}}{6}$$

2.5.3   CASE 3: STRAIN IN COMPRESSION STEEL REACHES ELASTIC LIMIT VALUE

Depth of neutral axis is given by

$$x_c^{(iii)} = d + \frac{\varepsilon_{s0}}{\phi_E} \tag{2.30}$$

By substituting Equation 2.30 in Equation 2.8, the expression for limit elastic curvature is obtained as

$$H_0 + H_1\,\phi_E + H_2\,\phi_E^2 + H_3\,\phi_E^3 = 0 \tag{2.31}$$

where the constants H_i (for i = 0 to 3) are given by

$$H_0 = \frac{b\varepsilon_{s0}^2(3\varepsilon_{c0} + \varepsilon_{s0})\sigma_{c0}}{3\varepsilon_{c0}^2}$$

$$H_1 = \frac{-P_E\varepsilon_{c0}^2 + b\varepsilon_{s0}[(D-d)E_s(p_c + p_t)\varepsilon_{c0}^2 + d\sigma_{c0}(2\varepsilon_{c0} - \varepsilon_{s0})]}{\varepsilon_{c0}^2} \tag{2.32}$$

$$H_2 = b\left[(3dD - D^2 - 2d^2)E_s p_t + \frac{d^2\sigma_{c0}(\varepsilon_{c0} - \varepsilon_{s0})}{\varepsilon_{c0}^2}\right]$$

$$H_3 = -\frac{bd^3\sigma_{c0}}{3\varepsilon_{c0}^2}$$

By solving Equation 2.31, only one real root (the second one) gives the limit elastic curvature as

$$\phi_E^{(iii)} = \frac{1}{12H_3}\left[-4\,H_2 - \frac{(2.5198 + 4.3645i)\left(H_2^2 - 3H_1\,H_3\right)}{\omega} - (1.5874 - 2.7495i)\omega\right] \tag{2.33}$$

where

$$\omega = \left[-2H_2^3 + 9H_1H_2H_3 - 27H_3^2H_0 + \sqrt{-4\left(H_2^2 - 3H_1H_3\right)^3 + \left(2H_2^3 - 9H_1H_2H_3 + 27H_3^2H_0\right)^2}\right]^{1/3} \tag{2.34}$$

By substituting Equation 2.33 in Equation 2.9, limit elastic bending moment can be obtained as follows:

$$M_E^{(iii)} = \frac{b}{2\varepsilon_{c0}^2}\left[\frac{M_1^{(iii)}}{\phi_E^{(iii)2}} + \frac{M_2^{(iii)}}{\phi_E^{(iii)}} + M_3^{(iii)} + M_4^{(iii)}\phi_E^{(iii)} + M_5^{(iii)}\phi_E^{(iii)2}\right] \tag{2.35}$$

where

$$M_1^{(iii)} = \frac{\varepsilon_{s0}^3(\varepsilon_{s0} - 4\varepsilon_{c0})\sigma_{c0}}{6}, \quad M_2^{(iii)} = \frac{(D-2d)(3\varepsilon_{c0} - \varepsilon_{s0})\varepsilon_{s0}^2\,\sigma_{c0}}{3},$$

$$M_3^{(iii)} = (d-D)\varepsilon_{s0}\left[(D-2d)E_s(p_t - p_c)\varepsilon_{c0}^2 - d(2\varepsilon_{c0} - \varepsilon_{s0})\sigma_{c0}\right],$$

$$(2.36)$$

$$M_4^{(iii)} = (D-d)(D-2d)^2 E_s p_t \varepsilon_{c0}^2 + \frac{d^2(3D-2d)(\varepsilon_{c0} - \varepsilon_{s0})\sigma_{c0}}{3},$$

$$M_5^{(iii)} = \frac{d^3(d-2D)\sigma_{c0}}{6}$$

2.5.4   CASE 4: STRAIN IN EXTREME COMPRESSION FIBER IN CONCRETE REACHES ELASTIC LIMIT VALUE

Now, the depth of neutral axis is given by

$$x_c^{(iv)} = \frac{\varepsilon_{c0}}{\phi_E}$$

$$(2.37)$$

By substituting Equation 2.37 in Equation 2.8, the expression for limit elastic curvature is obtained as

$$R_0 + R_1\,\phi_E + R_2\,\phi_E^2 = 0$$

$$(2.38)$$

where the constants R_i (for i = 0 to 2) are given by

$$R_0 = \frac{2b\varepsilon_{c0}\sigma_{c0}}{3}, \quad R_1 = -P_c + b(D-d)E_s\varepsilon_{c0}(p_c + p_t),$$

$$(2.39)$$

$$R_2 = -b(D-d)E_s[Dp_t - d(p_t - p_c)]$$

By solving Equation 2.38, the only real root (in this case, the first root) gives the limit elastic curvature as

$$\phi_E^{(iv)} = -\frac{R_1 + \sqrt{R_1^2 - 4R_0R_2}}{2R_2}$$

$$(2.40)$$

By substituting Equation 2.40 in Equation 2.9, limit elastic bending moment, M_E, can be obtained as follows:

$$M_E^{(iv)} = \frac{M_1^{(iv)}}{\phi_E^{(iv)2}} + \frac{M_2^{(iv)}}{\phi_E^{(iv)}} + M_3^{(iv)} + M_4^{(iv)}\phi_E^{(iv)}$$

$$(2.41)$$

where

$$M_1^{(iv)} = \frac{bD\varepsilon_{c0}\sigma_{c0}}{3}$$

$$M_2^{(iv)} = -\frac{1}{4} b\varepsilon_{c0}^2 \sigma_{c0}$$

$$M_3^{(iv)} = \frac{1}{2} b(D^2 + 2d^2 - 3dD)E_s(p_c - p_t)\varepsilon_{c0}$$ (2.42)

$$M_4^{(iv)} = \frac{1}{2} b(D^2 + 2d^2 - 3dD)E_s[Dp_t - d(p_c + p_t)]$$

It may be easily seen that for tension steel exceeding the maximum limit of 4%, as specified in many codes (e.g., see IS 456, 2000), case 4 will never result in a practical situation. For the case $(x_c > D)$, the limits of the integral in Equation 2.8 will be from $(0, D)$, which will also result in compression failure and hence are not discussed (compression failure is not of design interest for several disadvantages affiliated to such failure). Expressions for limit elastic moments are summarized below:

$$M_E = \begin{cases} M_E^{(ii)} & \text{if } p_t < p_{t,el} \\ M_E^{(iii)} & \text{if } p_t > p_{t,el} \end{cases}$$ (2.43)

where $p_{t,el}$, for two cases, namely, (1) axial force neglected, and (2) axial force considered, are given by the following equations:

$$P_{t,el} = p_c + \frac{D^2[D(3\varepsilon_{c0} - \varepsilon_{s0}) - 6d\varepsilon_{c0}]\sigma_{c0}}{6(D-d)(D-2d)^2 E_s \varepsilon_{c0}^2}$$ (2.44)

$$P_{t,el} = \frac{6(D-2d)^2\varepsilon_{c0}^2[P_E + b(d-D)E_s p_c \varepsilon_{s0}] + bD^2\varepsilon_{s0}[6d\varepsilon_{c0} + D(\varepsilon_{s0} - 3\varepsilon_{c0})]\sigma_{c0}}{6b(D-d)(D-2d)^2 E_s \varepsilon_{c0}^2 \varepsilon_{s0}}$$

(2.45)

2.6 PERCENTAGE OF STEEL FOR BALANCED SECTION

The percentage of reinforcement in tension and compression for balanced failure is obtained by considering both of the following conditions: (1) maximum compressive strain in concrete reaches ultimate limit strain and (2) strain in tensile reinforcement reaches ultimate limit. Balanced reinforcement for two cases is considered, namely, (1) for beams where axial force vanishes and (2) for beam/columns where

P-M interaction is predominantly present. For sections with vanishing axial force, depth of neutral axis is given by

$$x_c = \left(\frac{\varepsilon_{cu}}{\varepsilon_{cu} + \varepsilon_{su}} \right)(D - d) \qquad (2.46)$$

For vanishing axial force, the governing equation to determine the percentage of reinforcement is given by

$$P = \int_q^{x_c} b\sigma_c [\varepsilon_c(y)] dy + (A_{sc} - A_{st})\sigma_{s0} + qb\sigma_{c0} = 0 \qquad (2.47)$$

In explicit form, Equation 2.47 becomes

$$b(d - D)[\sigma_{c0}\varepsilon_{c0} - 3\varepsilon_{cu}\sigma_{c0} - 3(p_c - p_t)(\varepsilon_{cu} + \varepsilon_{su})\sigma_{s0}] = 0 \qquad (2.48)$$

By solving, the percentage of steel for a balanced section is obtained as

$$P_{t,bal} = P_c + \frac{(3\varepsilon_{cu} - \varepsilon_{c0})\sigma_{c0}}{3(\varepsilon_{cu} + \varepsilon_{su})\sigma_{s0}} \qquad (2.49)$$

For a known cross-section with a fixed percentage of compression reinforcement, Equation 2.49 gives the percentage of steel for a balanced section. It can be easily seen that for the assumed condition of strain in compression steel greater than elastic limit, Equation 2.49 shall yield the percentage of tension reinforcement for balanced sections whose overall depth exceeds 240 mm, which is a practical case of cross-section dimension of RC beams used in multistory building frames. For sections where axial force is predominantly present, the percentage of balanced reinforcement depends on the magnitude of axial force. By assuming the same hypothesis presented above, the depth of the neutral axis is given by Equation 2.46, but Equation 2.48 becomes as given below:

$$b(d - D)[\sigma_{c0}\varepsilon_{c0} - 3\varepsilon_{cu}\sigma_{c0} - 3(p_c - p_t)(\varepsilon_{cu} + \varepsilon_{su})\sigma_{s0}] = P_0 \qquad (2.50)$$

By solving, the percentage of steel for a balanced section is obtained as

$$P_{t,bal} = P_c + \frac{(3\varepsilon_{cu} - \varepsilon_{c0})\sigma_{c0}}{3(\varepsilon_{cu} + \varepsilon_{su})\sigma_{s0}} - \frac{P_0}{b(D - d)\sigma_{s0}} \qquad (2.51)$$

where P_0 is the axial force ($P_0 > 0$ if it is compression). For the known cross-section with a fixed percentage of compression reinforcement, Equation 2.51 gives the percentage of steel for a balanced section. In a similar manner, the percentage of compression reinforcement for a balanced section, by fixing p_t, can be obtained by inverting the relationship given in Equations 2.49 and 2.51 for respective axial force conditions.

2.7   ULTIMATE BENDING MOMENT-CURVATURE RELATIONSHIP

The study in this section is limited to RC sections imposed with tension failure, because the compression and balance failures do not have any practical significance in the displacement-based design approach, in particular. Let us consider two possible cases: (1) neutral axis position assumes negative values and (2) neutral axis position assumes positive values.

2.7.1   Neutral Axis Position Assuming Negative Values

By imposing the conditions $x_c = 0$ and $\phi = \varepsilon_{su}/(D-d)$ and solving Equation 2.2 with respect to P_t, for a specified range of $p_t < P_u + bd E_s p_c \varepsilon_{su}/b(d-D)\sigma_{s0}$, the depth of the neutral axis is given by

$$x_c = D - d - \frac{\varepsilon_{su}}{\phi_u} \quad \forall\, x_c < 0 \tag{2.52}$$

At collapse, the equilibrium equations become

$$P_u = -\sigma_{s0}A_{st} + \sigma_{sc}A_{sc} = b(d-D)[p_t\sigma_{s0} + E_s p_c(d-x_c)\phi_u] \tag{2.53}$$

$$M_u = (\sigma_{s0}A_{st} + \sigma_{sc}A_{sc})\left(\frac{D}{2} - d\right) = \frac{b(D-2d)}{2}(D-d)[p_t\sigma_{s0} + E_s p_c(x_c - d)\,\phi_u] \tag{2.54}$$

By solving Equation 2.53 with respect to ϕ_u, we obtain the ultimate curvature as

$$\phi_u = \frac{P_u + b(D-d)[\sigma_{s0}p_t + E_s p_c \varepsilon_{su}]}{bE_s p_c(D^2 + 2d^2 - 3dD)} \tag{2.55}$$

By substituting Equation 2.55 in Equation 2.54, the ultimate bending moment can be determined as

$$M_u = \frac{D-2d}{2}[P_u + 2b(D-d)p_t\sigma_{s0}] \tag{2.56}$$

It may be noted that the ultimate bending moment in this case is similar to one given by Equation 2.22 for elastic range.

2.7.2   Neutral Axis Position Assuming Positive Values

Under this condition at collapse, four different cases of tension failure of RC sections are possible, namely,

(a) $\varepsilon_{st} = \varepsilon_{su}$, $\varepsilon_{sc} < \varepsilon_{s0}$, $\varepsilon_{c,max} < \varepsilon_{c0}$,

(b) $\varepsilon_{st} = \varepsilon_{su}$, $\varepsilon_{sc} < \varepsilon_{s0}$, $\varepsilon_{c0} < \varepsilon_{c,max} < \varepsilon_{cu}$,

(c) $\varepsilon_{st} = \varepsilon_{su}$, $\varepsilon_{s0} < \varepsilon_{sc} < \varepsilon_{su}$, $\varepsilon_{c,max} < \varepsilon_{c0}$, (2.57)

(d) $\varepsilon_{st} = \varepsilon_{su}$, $\varepsilon_{s0} < \varepsilon_{sc} < \varepsilon_{cu}$, $\varepsilon_{c0} < \varepsilon_{c,max} < \varepsilon_{cu}$

As strain in tensile steel reaches its ultimate value causing tensile failure, in all the four cases mentioned above, the equation for computing the depth of the neutral axis (as a function of ultimate curvature) will remain unchanged and is given by

$$x_c^{(a-d)} = D - d - \frac{\varepsilon_{su}}{\phi_u}$$ (2.58)

Axial force and bending moment in the cross-section at collapse for case (a) are given by

$$P_u = \int_0^{x_c} b\sigma_c[\varepsilon_c(y)]\, dy - \sigma_{s0}A_{st} + \sigma_{sc}A_{sc}$$ (2.59)

$$M_u = \int_0^{x_c} b\sigma_c[\varepsilon_c(y)]\left(\frac{D}{2} - y\right)dy + (\sigma_{s0}A_{st} + \sigma_{sc}A_{sc})\left(\frac{D}{2} - d\right)$$ (2.60)

By substituting Equation 2.58 in Equation 2.59 we get

$$J_0 + J_1\phi_u + J_2\phi_u^2 + J_3\phi_u^3 = 0$$ (2.61)

where the constants $J_{i=0,1,2,3}$ are given by

$$J_0 = \frac{b\varepsilon_{su}^2(3\varepsilon_{c0} + \varepsilon_{su})\sigma_{c0}}{3\varepsilon_{c0}^2},$$

$$J_1 = \frac{-P_u\varepsilon_{c0}^2 + b(d-D)\left[\left(E_s p_c\varepsilon_{c0}^2 + \sigma_{c0}(2\varepsilon_{c0} + \varepsilon_{su})\right)\varepsilon_{su} + p_t\sigma_{s0}\varepsilon_{c0}^2\right]}{\varepsilon_{c0}^2},$$

(2.62)

$$J_2 = b\left[(D^2 + 2d^2 - 3dD)E_s p_c + \frac{(d-D)^2(\varepsilon_{c0} + \varepsilon_{su})\sigma_{c0}}{\varepsilon_{c0}^2}\right],$$

$$J_3 = \frac{b(d-D)^3\sigma_{c0}}{3\varepsilon_{c0}^2}$$

By solving Equation 2.61, the real root (in this case, the third root) gives the ultimate curvature as

$$\phi_u^{(a)} = \frac{1}{12J_3}\left[-4\,J_2 - \frac{(2.5198 - 4.3645i)\left(J_2^2 - 3J_1 J_3\right)}{\alpha} - (1.5874 + 2.7495i)\alpha\right]$$

(2.63)

where

$$\alpha = \left[-2J_2^3 + 9J_1J_2J_3 - 27J_3^2J_0 + \sqrt{-4\left(J_2^2 - 3J_1J_3\right)^3 + \left(2J_2^3 - 9J_1J_2J_3 + 27J_3^2J_0\right)^2} \right]^{1/3}$$

(2.64)

By substituting Equation 2.63 in Equation 2.60, ultimate moment is given by

$$M_u^{(a)} = \frac{b}{2\varepsilon_{c0}^2}\left[\frac{M_1^{(a)}}{\phi_u^{(a)2}} + \frac{M_2^{(a)}}{\phi_u^{(a)}} + M_3^{(a)} + M_4^{(a)}\phi_u^{(a)} + M_5^{(a)}\phi_u^{(a)2} \right]$$

(2.65)

where the superscript $^{(a)}$ stands for the case (a); the constants of the above equation are given by

$$M_1^{(a)} = \frac{\varepsilon_{su}^3(4\varepsilon_{c0} + \varepsilon_{su})\sigma_{c0}}{6}, \quad M_2^{(a)} = \frac{(2d - D)\varepsilon_{su}^2(3\varepsilon_{c0} + \varepsilon_{su})\sigma_{c0}}{3},$$

$$M_3^{(a)} = (D - d)\left[(2d - D)E_s p_c \varepsilon_{su}\varepsilon_{c0}^2 - d\varepsilon_{su}(2\varepsilon_{c0} + \varepsilon_{su})\sigma_{c0} + (D - 2d)p_t\varepsilon_{c0}^2\sigma_{s0} \right],$$

$$M_4^{(a)} = \frac{(D - d)\left[-3(D - 2d)^2 E_s p_c \varepsilon_{c0}^2 + (d - D)(2d + D)(\varepsilon_{c0} + \varepsilon_{su})\sigma_{c0} \right]}{3},$$

(2.66)

$$M_5^{(a)} = \frac{(d - D)^3(D + d)\sigma_{c0}}{6}$$

Axial force and bending moment in the cross-section at collapse for case (b) are given by

$$P_u = \int_q^{x_c} b\sigma_c[\varepsilon_c(y)]\, dy - A_{st}\sigma_{s0} + A_{sc}\sigma_{sc} + q\, b\sigma_{c0}$$

(2.67)

$$M_u = \int_q^{x_c} b\sigma_c[\varepsilon_c(y)]\left(\frac{D}{2} - y \right) dy + (A_{st}\sigma_{s0} + A_{sc}\sigma_{sc})\left(\frac{D}{2} - d \right) + \frac{q\, b\sigma_{c0}}{2}(D - q)$$

(2.68)

By substituting the Equation 2.58 in 2.67, we get

$$Q_0 + Q_1\phi_u + Q_2\phi_u^2 = 0$$

(2.69)

where the constants $Q_{i = 0,1,2}$ are given by

$$Q_0 = -\frac{b\sigma_{c0}(\varepsilon_{c0} + 3\varepsilon_{su})}{3},$$

$$Q_1 = b(D - d)(\sigma_{c0} - E_s p_c \varepsilon_{su} + \sigma_{s0}p_t) - P_u,$$

(2.70)

$$Q_2 = bE_s p_c(D^2 + 2d^2 - 3dD)$$

By solving Equation 2.69, the first root of the quadratic, representing the ultimate curvature, is given as

$$\phi_u^{(b)} = \frac{-Q_1 + \sqrt{Q_1^2 - 4Q_0 Q_2}}{2Q_2} \tag{2.71}$$

By substituting Equation 2.71 in Equation 2.68, ultimate moment is obtained as

$$M_u^{(b)} = \frac{b}{2}\left[\frac{M_1^{(b)}}{\phi_u^{(b)2}} + \frac{M_2^{(b)}}{\phi_u^{(b)}} + M_3^{(b)} + M_4^{(b)}\phi_u^{(b)}\right] \tag{2.72}$$

where

$$M_1^{(b)} = \frac{\left(\varepsilon_{c0}^2 + 4\varepsilon_{c0}\varepsilon_{su} + 6\varepsilon_{su}^2\right)\sigma_{c0}}{6},$$

$$M_2^{(b)} = \frac{(D - 2d)(\varepsilon_{c0} + 3\varepsilon_{su})\sigma_{c0}}{3}, \tag{2.73}$$

$$M_3^{(b)} = (D - d)[d(2E_s p_c \varepsilon_{su} + \sigma_{c0} - 2p_t\sigma_{s0}) + D(p_t\sigma_{s0} - E_s p_c \varepsilon_{su})]$$

$$M_4^{(b)} = (D - d)(D - 2d)^2 E_s p_c$$

Axial force and bending moment in the cross-section at collapse for case (c) are given by

$$P_u = \int_0^{x_c} b\sigma_c[\varepsilon_c(y)]\,dy + \sigma_{s0}(A_{sc} - A_{st}) \tag{2.74}$$

$$M_u = \int_0^{x_c} b\sigma_c[\varepsilon_c(y)]\left(\frac{D}{2} - y\right)dy + (A_{st} + A_{sc})\sigma_{s0}\left(\frac{D}{2} - d\right) \tag{2.75}$$

By substituting Equation 2.58 in Equation 2.74, we get

$$W_0 + W_1\phi_u + W_2\phi_u^2 + W_3\phi_u^3 = 0 \tag{2.76}$$

where the constants $W_{i=0,1,2,3}$ are given by

$$W_0 = J_0,$$

$$W_1 = \frac{-P_u\varepsilon_{c0}^2 + b(d - D)[(2\varepsilon_{c0} + \varepsilon_{su})\sigma_{c0}\varepsilon_{su} + (p_t - p_c)\sigma_{s0}\varepsilon_{c0}^2]}{\varepsilon_{c0}^2},$$

$$W_2 = \frac{b(d - D)^2\sigma_{c0}(\varepsilon_{c0} + \varepsilon_{su})}{\varepsilon_{c0}^2}, \tag{2.77}$$

$$W_3 = J_3$$

where $J_{0.3}$ can be seen from Equation 2.62. By solving Equation 2.76, the real root (in this case, it is the third root) gives the ultimate curvature as

$$\phi_u^{(c)} = \frac{1}{12W_3}\left[-4\ W_2 - \frac{(2.5198 - 4.3645i)\left(W_2^2 - 3W_1\ W_3\right)}{\beta} - (1.5874 + 2.7495i)\beta\right]$$

(2.78)

where

$$\beta = \left[-2W_2^3 + 9W_1W_2W_3 - 27W_3^2W_0 + \sqrt{-4\left(W_2^2 - 3W_1W_3\right)^3 + \left(2W_2^3 - 9W_1W_2W_3 + 27W_3^2W_0\right)^2}\right]^{1/3}$$

(2.79)

By substituting Equation 2.78 in Equation 2.74, ultimate moment is obtained as

$$M_u^{(c)} = \frac{b}{2\varepsilon_{c0}^2}\left[\frac{M_1^{(c)}}{\phi_u^{(c)2}} + \frac{M_2^{(c)}}{\phi_u^{(c)}} + M_3^{(c)} + M_4^{(c)}\phi_u^{(c)} + M_5^{(c)}\phi_u^{(c)2}\right]$$

(2.80)

where

$$M_3^{(c)} = d(d - D)\varepsilon_{su}\,\sigma_{c0}\,(2\varepsilon_{c0} + \varepsilon_{su}) + (D^2 + 2d^2 - 3dD)(p_c + p_t)\sigma_{s0}\,\varepsilon_{c0}^2,$$

$$M_4^{(c)} = \frac{(d - D)^2(2d + D)(\varepsilon_{c0} + \varepsilon_{su})\sigma_{c0}}{3},$$

(2.81)

$$M_1^{(c)} = M_1^{(a)},\ \ M_2^{(c)} = M_2^{(a)},\ \ M_5^{(c)} = M_5^{(a)}$$

Axial force and bending moment in the cross-section at collapse for case (d) are given by

$$P_u = \int_q^{x_c} b\sigma_c[\varepsilon_c(y)]dy + (A_{sc} - A_{st})\sigma_{s0} + qb\sigma_{c0}$$

(2.82)

$$M_u = \int_q^{x_c} b\sigma_c[\varepsilon_c(y)]\left(\frac{D}{2} - y\right)dy + (A_{st} + A_{sc})\sigma_{s0}\left(\frac{D}{2} - d\right) + \frac{qb\sigma_{c0}}{2}(D - q)$$

(2.83)

By substituting Equation 2.58 in Equation 2.82 and solving, the ultimate curvature is obtained as

$$\phi_u^{(d)} = \frac{b\sigma_{c0}(\varepsilon_{c0} + 3\varepsilon_{su})}{3[b(D - d)(\sigma_{c0} + \sigma_{s0}(p_c - p_t) - P_u)]}$$

(2.84)

By substituting Equation 2.84 in Equation 2.83, the ultimate bending moment is obtained as

$$M_u^{(d)} = \frac{b\sigma_{c0}}{12} \left[\begin{array}{l} 6(D-d)\left(d+(D-2d)\,(p_c+p_t)\dfrac{\sigma_{s0}}{\sigma_{c0}}\right) + \\[2ex] \dfrac{2(D-2d)(\varepsilon_{c0}+3\varepsilon_{su})}{\phi_u^{(iv)}} - \dfrac{\varepsilon_{c0}^2+4\varepsilon_{c0}\varepsilon_{su}+6\varepsilon_{su}^2}{\phi_u^{(iv)2}} \end{array} \right] \tag{2.85}$$

For the condition of $D < \frac{d(2\varepsilon_{c0}-\varepsilon_{s0}+\varepsilon_{su})}{\varepsilon_{c0}-\varepsilon_{s0}}$, ultimate moment, derived above, takes the following form:

$$M_u = \begin{cases} M_u^{(a)} & \text{if } p_t < p_t^{(1)} \\ M_u^{(b)} & \text{if } p_t^{(1)} < p_t < p_t^{(2)} \\ M_u^{(d)} & \text{if } p_t^{(2)} < p_t \end{cases} \tag{2.86}$$

where

$$p_t^{(1)} = \frac{3(\varepsilon_{c0}+\varepsilon_{su})[P_u+bE_sP_c(d(2\varepsilon_{c0}+\varepsilon_{su})-D\varepsilon_{c0})]+2b(d-D)\varepsilon_{c0}\sigma_{c0}}{3b(d-D)(\varepsilon_{c0}+\varepsilon_{su})\sigma_{s0}},$$

$$p_t^{(2)} = \frac{3(\varepsilon_{s0}+\varepsilon_{su})[P_u+b(d-D)E_sp_c\varepsilon_{s0}]+b\sigma_{c0}[D(\varepsilon_{c0}-3\varepsilon_{s0})+d(3\varepsilon_{s0}-2\varepsilon_{c0}-3\varepsilon_{su})]}{3b(d-D)(\varepsilon_{s0}+\varepsilon_{su})\sigma_{s0}}$$

$$\tag{2.87}$$

The percentage of tension reinforcements is determined by imposing the following conditions:

1. $p_t^{(1)}$ is determined by imposing $\varepsilon_{st} = \varepsilon_{su}$, $\varepsilon_{c,max} = \varepsilon_{c0}$ and solving Equation 2.60 with respect to p_t.
2. $p_t^{(2)}$ is determined by imposing the $\varepsilon_{st} = \varepsilon_{su}$, $\varepsilon_{sc} = \varepsilon_{s0}$ and solving Equation 2.68 with respect to p_t.

For the other condition, namely, $D > \frac{d(2\varepsilon_{c0}-\varepsilon_{s0}+\varepsilon_{su})}{\varepsilon_{c0}-\varepsilon_{s0}}$, ultimate moment now takes a different form as given below:

$$M_u = \begin{cases} M_u^{(a)} & \text{if } p_t < p_t^{(3)} \\ M_u^{(c)} & \text{if } p_t^{(3)} < p_t < p_t^{(4)} \\ M_u^{(d)} & \text{if } p_t^{(4)} < p_t \end{cases} \tag{2.88}$$

where

$$p_t^{(3)} = p_c + \frac{3P_u(\varepsilon_{c0} + \varepsilon_{su}) + 2b(d - D)\varepsilon_{c0}\sigma_{c0}}{3b(d - D)(\varepsilon_{c0} + \varepsilon_{su})\sigma_{s0}},$$

$$p_t^{(4)} = p_c + \frac{P_u}{b(d - D)\sigma_{s0}} + \frac{[D\varepsilon_{s0} + d(\varepsilon_{su} - \varepsilon_{s0})]^2[D(\varepsilon_{c0} - 3\varepsilon_{s0}) + d(6\varepsilon_{c0} - \varepsilon_{s0} + \varepsilon_{su})]\sigma_{c0}}{3\varepsilon_{c0}^2(\varepsilon_{s0} + \varepsilon_{su})(D - 2d)^2(d - D)\sigma_{s0}} \quad (2.89)$$

The percentage of tension reinforcements is determined by imposing the following conditions:

1. $p_t^{(3)}$ is determined by imposing the $\varepsilon_{st} = \varepsilon_{su}$, $\varepsilon_{c,max} = \varepsilon_{c0}$ and solving Equation 2.75 with respect to p_t;

2. $p_t^{(4)}$ is determined by imposing the $\varepsilon_{st} = \varepsilon_{su}$, $\varepsilon_{sc} = \varepsilon_{s0}$ and solving Equation 2.83 respect to p_t.

For the condition $D = \frac{d(2\varepsilon_{c0} - \varepsilon_{s0} + \varepsilon_{su})}{\varepsilon_{c0} - \varepsilon_{s0}}$, ultimate moment is given by

$$M_u = \begin{cases} M_u^{(a)} & \text{if } p_t < p_t^* \\ M_u^{(d)} & \text{if } p_t^* < p_t \end{cases} \qquad p_t^* = p_t^{(1)} = p_t^{(2)} = p_t^{(3)} = p_t^{(4)} \qquad (2.90)$$

2.8 NUMERICAL STUDIES AND DISCUSSIONS

An example RC section of 300×500 is considered for the study. The section is reinforced on both tension and compression zones whose percentage is varied to study their influence on the curvature ductility. Concrete with compressive cube strength of 30 N/mm² and steel with yield strength of 415 N/mm² are considered. Figure 2.2 shows the variation of elastic moment with percentage of tension reinforcement for a constant compression reinforcement consisting of 4Φ22. It is seen that the limit elastic moment increases linearly for the case of tensile steel reaching its yield limit while strain in concrete is within the elastic limit (see the curve governed by Equations 2.22 and 2.28). For other cases, namely, (1) strain in compression steel reaches elastic limit (see the curve governed by Equation 2.35 as well as (2) crushing failure where strain in extreme fiber in concrete reaches elastic limit (see the curve governed by Equation 2.41), the influence of the percentage of tension reinforcement on the limit elastic moment is marginal. Although there is a sharp rise for lower percentages of reinforcements, this increase becomes marginal for higher percentage values. The point of intersection of moment profiles governed by Equations 2.22 and 2.28 with that of Equation 2.35 gives the limit value of the percentage of tensile reinforcement ($p_{t,elastic}$). A percentage of tensile steel less than this value results in yielding of tensile steel while greater values result in yielding of compression steel. The point of intersection of moment profiles governed by Equations 2.22 and 2.28 with that of Equation 2.41 is not of significant importance because the latter results in crushing failure of concrete. It is evident that the percentage of tensile reinforcement influences limit elastic moment considerably

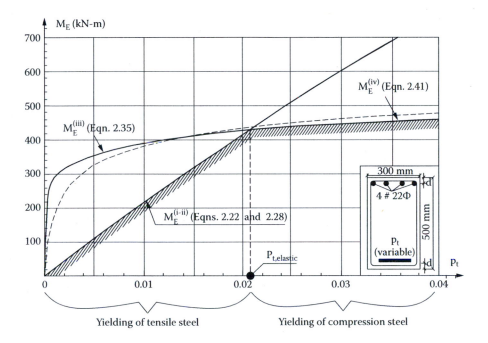

FIGURE 2.2 Variation of elastic moment with percentage of tensile steel reinforcement.

in the case of ductile failure only. It may be noted that Figure 2.2 plots the moment variation based on the same governing equations used subsequently for estimating the moment-curvature relationship. It can also be seen that limit elastic moment is given by the minimum of the four values given by the Equations 2.22, 2.28, 2.35, and 2.41, respectively. The trace of the point along the hatched line gives the minimum limit elastic moment, thus obtained. Figure 2.3 shows the moment-curvature plots for the RC section reinforced with 4#22Φ on tension face but varying the compression steel. It can be seen from the figure that for a fixed percentage of tensile reinforcement, influence of variation of compression reinforcement on moment-curvature is only marginal. Also, there exists at least one critical value of the percentage of both tensile and compression reinforcement that reduces the curvature ductility to the minimum. The proposed analytical expressions are capable of tracing this critical value, so that it can be avoided for a successful design of the section.

The effect of axial force on moment-curvature is also studied by subjecting the RC section reinforced with 4#22Φ, both on compression and tension zones. The section is subjected to compressive axial force only as the tensile force limits the curvature and cannot be helpful in predicting the desired behavior. Figure 2.4 presents the moment-curvature for different axial forces considered. Moment-curvatures seen in the figure show linear response in elastic range and hardening-like response in elastic-plastic range. The figure shows that increase in axial force results in increase of moments and curvature at both the elastic and ultimate levels. For all four cases shown in the figure, there is only a marginal increase in ultimate moment with respect to their corresponding limit elastic moment. For the numerical cases examined, it is

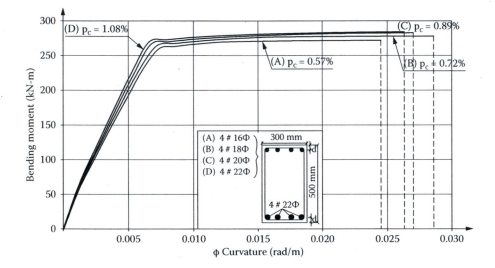

FIGURE 2.3 Variation of moment-curvature with percentage of compression reinforcement.

therefore stated that the variation in magnitude of axial force does not influence the ductility ratio much in comparison to its influence on limit elastic and ultimate moments; however, higher axial forces tend to reduce the curvature ductility. The critical value of axial force, beyond which a reduction is caused in curvature ductility, can also be obtained from the proposed analytical hypothesis.

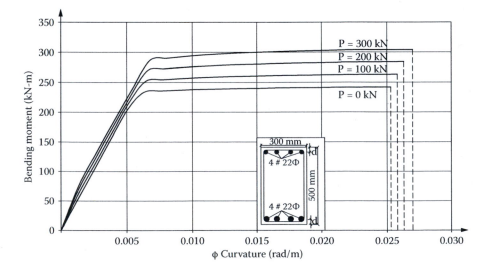

FIGURE 2.4 Moment-curvature relationship for different axial forces.

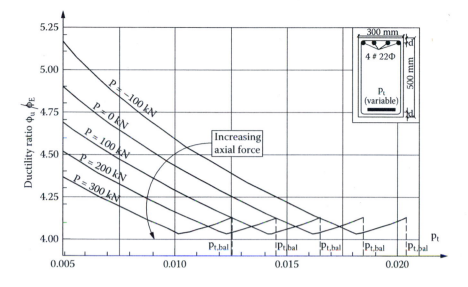

FIGURE 2.5 Variation of curvature ductility with percentage of tensile steel reinforcement.

The influence of the percentage of reinforcing steel on ductility ratio for different axial forces is also studied by examining two cases:(1) by varying steel percentage in tension, with 4#22Φ on the compression side; and (2) by varying the percentage of compression reinforcement, with 4#22Φ on the tension side. Figures 2.5 and 2.6 show the influence of tensile and compression reinforcement on curvature ductility, respectively. Figure 2.5 shows that plastic softening behavior is observed in the section under large curvature amplitudes. This may be attributed to the expected failure pattern (local collapse mechanism) of the structural members of building frames located in seismic areas. Larger ductility ratios for reduced tensile reinforcement prompts the design of members initiating ductile failure. However, tensile reinforcement closer to $p_{t,bal}$ will result in more curvature ductility since there is a marginal reduction seen due to the kink in the curve for (lesser) values closer to $p_{t,bal}$. Figure 2.6 shows that maximum curvature ductility is obtained for compression reinforcement equal to $p_{c,bal}$, when the section is subjected to axial compressive force. However, for tensile axial forces, the same percentage of compression steel as of tension steel ($p_c = p_t$) gives the maximum curvature ductility. It can be therefore summarized that the percentage of tension reinforcement influences curvature ductility to a larger extent and therefore demands good ductile detailing in the members of building frames located in seismic areas. Focus should be on this aspect while designing structures in seismic areas. Studies conducted by researchers with respect to the recent development in codes in this aspect (for example, see Amador and Nadyane 2008) also verified the same for a safe distribution of earthquake forces without complete collapse of the building. A spreadsheet program is used to estimate the moment-curvature relationship by iteration, after simplifying the complexities involved in such an estimate. The values are estimated in two ranges, namely, (1) elastic and (2) elastic-plastic,

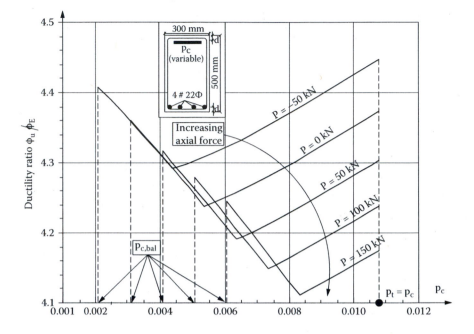

FIGURE 2.6 Variation of curvature ductility with percentage of compression steel reinforcement.

separately. The results given in the closed form are useful for researchers but not as much for practicing engineers; a simplified spreadsheet program is prepared to facilitate ready use for practicing engineers.

With the spreadsheet program, moment-curvature relationship of the RC section, reinforced with 4#22Φ, both in tension and compression sides, is now plotted for different axial loads (only compressive). The curves are compared with those obtained by using the proposed analytical expressions. Figure 2.7 shows the comparison of the curves obtained by employing both numerical and analytical procedures. By comparing, it can be seen that there is practically no difference between the curves in the elastic range, whereas there exists a marginal difference in the plastic range. However, both procedures estimate the same ultimate curvature and the ultimate moments as well. Also, the curvature ductility ratio obtained by both procedures remains the same. With regard to their close agreement, the proposed closed-form expressions for moment-curvature relationship, accounting for nonlinear characteristics of constitutive materials according to Eurocode, are thus qualified for use in seismic design and in structural assessments as well. A detailed procedure to obtain the moment-curvature relationship using the spreadsheet program is presented in Section 2.10. For easy reference to practicing engineers, Figures 2.8 to 2.15 show the moment-curvature plots for a few RC sections used in common practice; used are the relevant percentage of tensile and compression reinforcement. Please note that these are plotted for pure bending case only

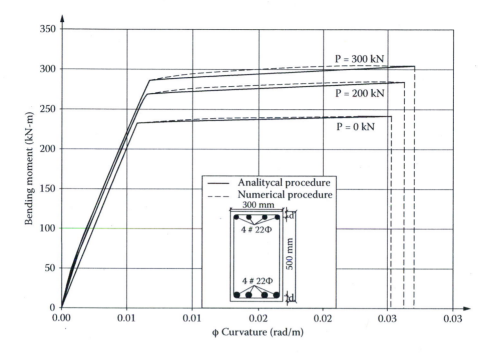

FIGURE 2.7 Comparison of moment-curvature by analytical and numerical procedures.

(P = 0). Tables 2.1 to 2.14 show the values for M – Φ relationship for the relevant sections considered in the analysis.

It can be inferred from the above discussions that a detailed trace of moment-curvature relationship is inevitable for successful seismic design of structures. The relationship is, however, very complex as a result of many factors: (1) constitutive material's nonlinear response; (2) magnitude of axial load and their nature; as well as (3) cross-section properties and percentage of reinforcement (tensile steel, in particular). Numerical studies conducted lead to useful design guidelines of multistory RC buildings. The upper-floor elements (beams, in particular) are designed to have ductile failure, which in turn permits large curvature ductility. This, in fact, helps the formation of plastic hinges at upper floors (on beams, in particular with a strong column–weak beam design concept) first, enabling effective redistribution of moments; this subsequently enables the formation of plastic hinges at lower floors. On the contrary, a column member, usually subjected to larger axial force, is designed without much increase in compression reinforcement because this does not help to improve its curvature ductility. However, in building frames under seismic loads, columns reinforced on two sides only will either be in tension or in compression, and hence $p_t = p_c$ holds well.

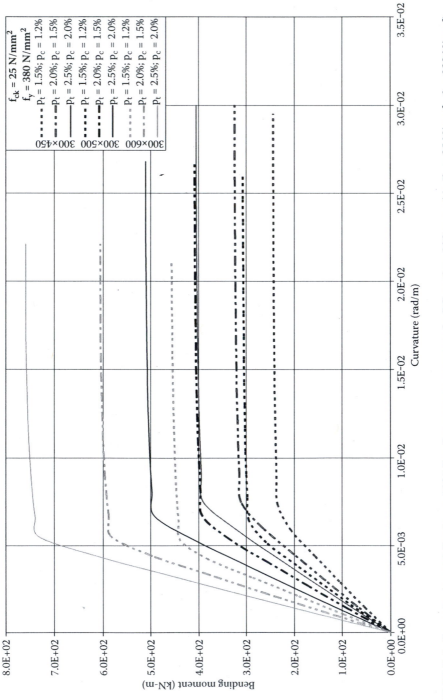

FIGURE 2.8 (See color insert following p. 138.) Bending moment-curvature for RC sections 300 mm wide ($f_{ck} = 25$ N/mm², $f_y = 380$ N/mm²).

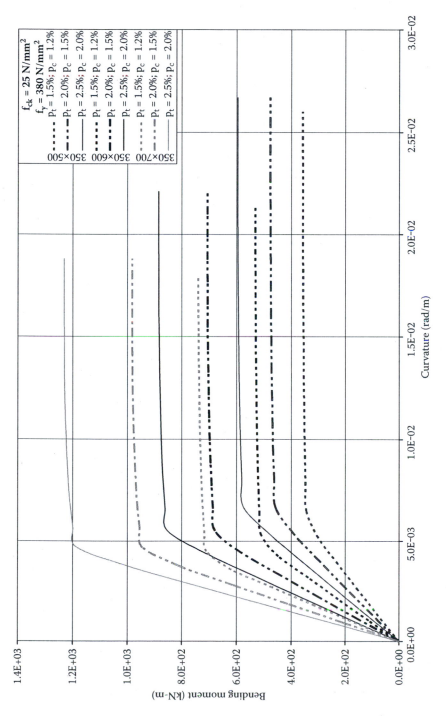

FIGURE 2.9 (See color insert following p. 138.) Bending moment-curvature for RC sections 350 mm wide ($f_{ck} = 25$ N/mm^2, $f_y = 380$ N/mm^2).

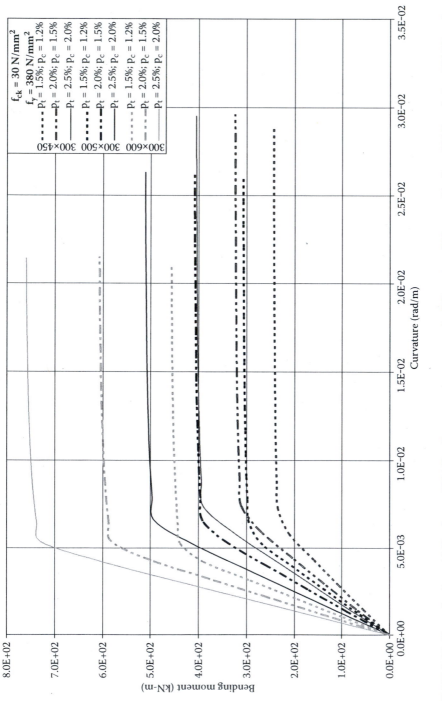

FIGURE 2.10 (See color insert following p. 138.) Bending moment-curvature for RC sections 300 mm wide ($f_{ck} = 30$ N/mm², $f_y = 380$ N/mm²).

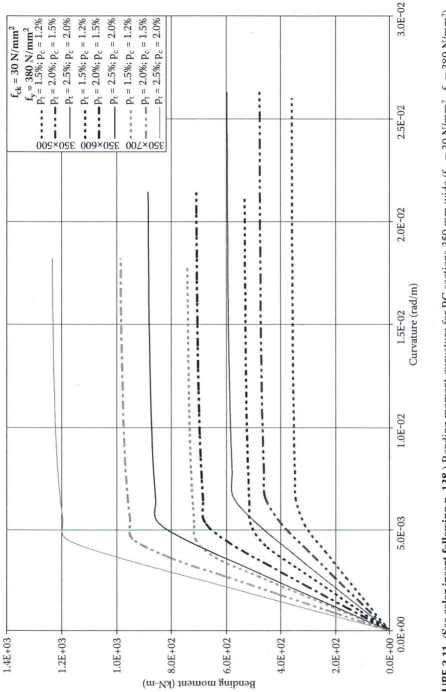

FIGURE 2.11 (See color insert following p. 138.) Bending moment-curvature for RC sections 350 mm wide ($f_{ck} = 30$ N/mm², $f_y = 380$ N/mm²).

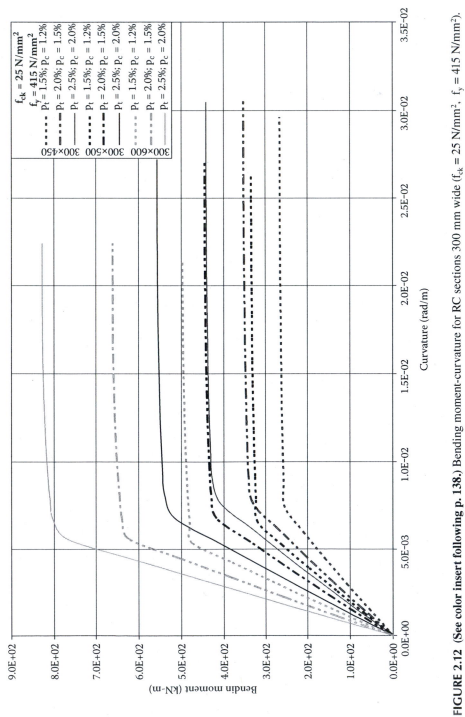

FIGURE 2.12 (See color insert following p. 138.) Bending moment-curvature for RC sections 300 mm wide (f_{ck} = 25 N/mm², f_y = 415 N/mm²).

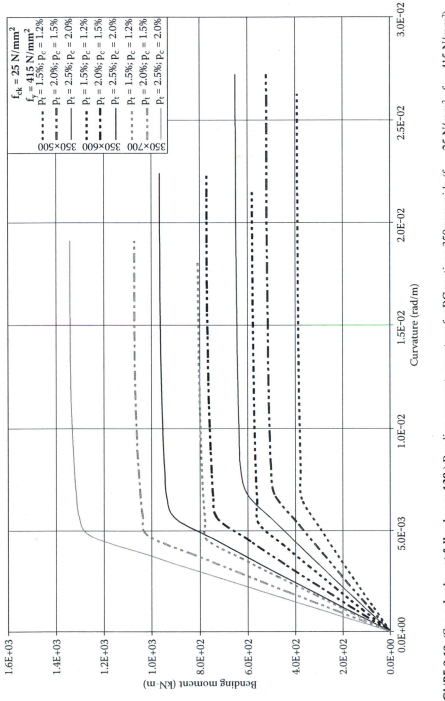

FIGURE 2.13 (**See color insert following p. 138.**) Bending moment-curvature for RC sections 350 mm wide ($f_{ck} = 25$ N/mm², $f_y = 415$ N/mm²).

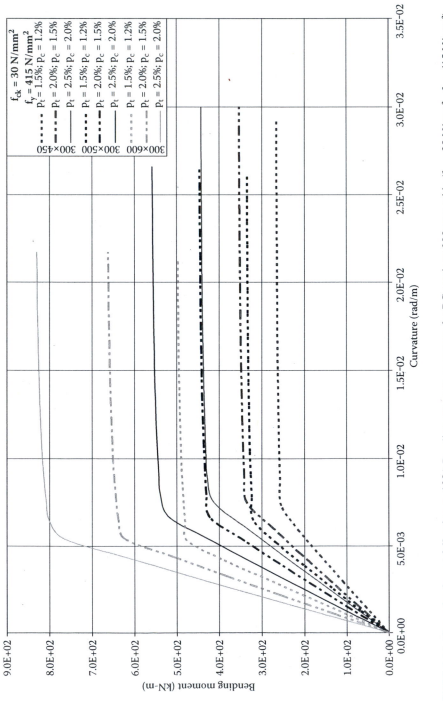

FIGURE 2.14 (See color insert following p. 138.) Bending moment-curvature for RC sections 300 mm wide (f_{ck} = 30 N/mm², f_y = 415 N/mm²).

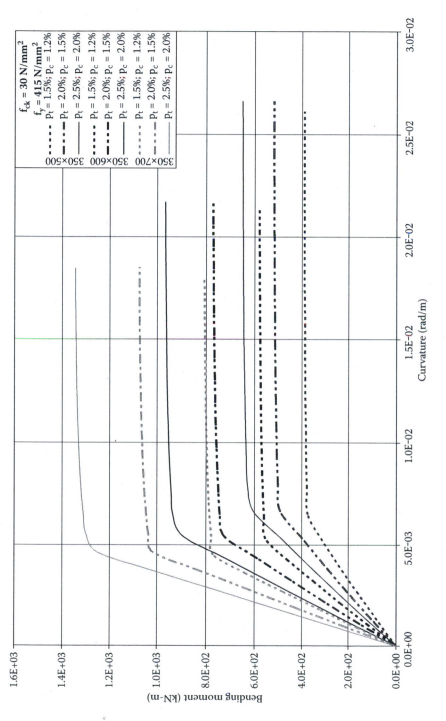

FIGURE 2.15 (See color insert following p. 138.) Bending moment-curvature for RC sections 350 mm wide (f_{ck} = 30 N/mm², f_y = 415 N/mm²).

TABLE 2.1
M-Φ for RC Section 300 × 450 (p_t = 1.5%, p_c = 1.2%, f_{ck} = 25 N/mm^2, f_y = 380 N/mm^2)

x_c	$\varepsilon_{c,max}$	$\sigma_{c,max}$	ε_{st}	σ_{st}	ε_{sc}	σ_{sc}	Φ	M
(mm)		(kN/m^2)		(kN/m^2)		(kN/m^2)	(rad/m)	(kN-m)
175.45	0.000002	0.019	0.000002	0.514	0.000001	0.305	0.000010	0.364
177.18	0.000261	2.693	0.000358	75.215	0.000217	45.588	0.001475	53.264
179.02	0.000528	5.053	0.000711	149.287	0.000440	92.318	0.002950	105.604
180.98	0.000801	7.060	0.001058	222.111	0.000668	140.296	0.004425	156.938
183.06	0.001080	8.691	0.001398	293.563	0.000903	189.647	0.005900	207.171
173.44	0.001279	9.591	0.001818	330.435	0.001058	222.152	0.007375	234.406
155.36	0.001375	9.947	0.002342	330.435	0.001109	232.977	0.008850	236.584
141.56	0.001462	10.225	0.002875	330.435	0.001152	241.891	0.010325	238.133
130.64	0.001542	10.444	0.003414	330.435	0.001188	249.389	0.011800	239.280
121.76	0.001616	10.618	0.003959	330.435	0.001218	255.799	0.013275	240.157
114.38	0.001687	10.754	0.004508	330.435	0.001245	261.357	0.014750	240.845
108.14	0.001755	10.857	0.005060	330.435	0.001268	266.231	0.016225	241.395
102.79	0.001819	10.933	0.005615	330.435	0.001288	270.549	0.017700	241.842
98.15	0.001882	10.985	0.006172	330.435	0.001307	274.407	0.019175	242.211
94.08	0.001943	11.014	0.006730	330.435	0.001323	277.883	0.020650	242.518
90.49	0.002002	11.023	0.007290	330.435	0.001338	281.036	0.022125	242.777
87.29	0.002060	11.023	0.007852	330.435	0.001352	283.915	0.023600	242.998
84.42	0.002117	11.023	0.008415	330.435	0.001365	286.555	0.025075	243.186
81.83	0.002173	11.023	0.008978	330.435	0.001376	288.984	0.026550	243.350
79.48	0.002228	11.023	0.009543	330.435	0.001387	291.226	0.028025	243.492
77.35	0.002282	11.023	0.010108	330.435	0.001397	293.303	0.029500	243.617

TABLE 2.2
M-Φ for RC Section 300 × 500 (p_t = 1.5%, p_c = 1.2%, f_{ck} = 25 N/mm^2, f_y = 380 N/mm^2)

x_c	$\varepsilon_{c,max}$	$\sigma_{c,max}$	ε_{st}	σ_{st}	ε_{sc}	σ_{sc}	Φ	M
(mm)		(kN/m^2)		(kN/m^2)		(kN/m^2)	(rad/m)	(kN-m)
195.46	0.000002	0.022	0.000003	0.577	0.000002	0.347	0.000010	0.514
197.34	0.000257	2.647	0.000354	74.436	0.000218	45.684	0.001300	66.344
199.35	0.000518	4.973	0.000704	147.775	0.000440	92.465	0.002600	131.584
201.48	0.000786	6.960	0.001047	219.918	0.000669	140.442	0.003900	195.626
203.75	0.001059	8.586	0.001385	290.749	0.000903	189.731	0.005200	258.357
194.12	0.001262	9.522	0.001793	330.435	0.001067	224.020	0.006500	295.025
173.58	0.001354	9.873	0.002312	330.435	0.001120	235.192	0.007800	297.850
157.90	0.001437	10.150	0.002840	330.435	0.001164	244.417	0.009100	299.863
145.47	0.001513	10.370	0.003375	330.435	0.001201	252.194	0.010400	301.358
135.35	0.001584	10.546	0.003915	330.435	0.001233	258.855	0.011700	302.503

(*Continued*)

TABLE 2.2 (CONTINUED)
M-Φ for RC Section 300×500 ($p_t = 1.5\%$, $p_c = 1.2\%$, $f_{ck} = 25$ N/mm², $f_y = 380$ N/mm²)

x_c (mm)	$\varepsilon_{c,max}$	$\sigma_{c,max}$ (kN/m²)	ε_{st}	σ_{st} (kN/m²)	ε_{sc}	σ_{sc} (kN/m²)	Φ (rad/m)	M (kN-m)
126.94	0.001650	10.686	0.004460	330.435	0.001260	264.638	0.013000	303.403
119.82	0.001713	10.797	0.005008	330.435	0.001284	269.715	0.014300	304.125
113.70	0.001774	10.882	0.005558	330.435	0.001306	274.214	0.015600	304.712
108.40	0.001832	10.946	0.006111	330.435	0.001325	278.234	0.016900	305.198
103.75	0.001888	10.989	0.006666	330.435	0.001342	281.854	0.018200	305.604
99.63	0.001943	11.014	0.007222	330.435	0.001358	285.135	0.019500	305.947
95.96	0.001996	11.023	0.007780	330.435	0.001372	288.127	0.020800	306.240
92.67	0.002048	11.023	0.008339	330.435	0.001385	290.871	0.022100	306.491
89.71	0.002099	11.023	0.008899	330.435	0.001397	293.396	0.023400	306.708
87.01	0.002149	11.023	0.009460	330.435	0.001408	295.728	0.024700	306.898
84.56	0.002199	11.023	0.010021	330.435	0.001419	297.889	0.026000	307.065

TABLE 2.3
M-Φ for RC Section 300×600 ($p_t = 1.5\%$, $p_c = 1.2\%$, $f_{ck} = 25$ N/mm², $f_y = 380$ N/mm²)

x_c (mm)	$\varepsilon_{c,max}$	$\sigma_{c,max}$ (kN/m²)	ε_{st}	σ_{st} (kN/m²)	ε_{sc}	σ_{sc} (kN/m²)	Φ (rad/m)	M (kN-m)
235.47	0.000002	0.026	0.000003	0.703	0.000002	0.431	0.000010	0.929
237.69	0.000252	2.602	0.000352	73.971	0.000220	46.233	0.001060	97.705
240.06	0.000509	4.896	0.000699	146.890	0.000445	93.518	0.002120	193.868
242.56	0.000771	6.863	0.001041	218.664	0.000676	141.948	0.003180	288.358
245.22	0.001040	8.482	0.001377	289.184	0.000913	191.632	0.004240	381.023
234.19	0.001241	9.437	0.001780	330.435	0.001082	227.265	0.005300	437.422
208.89	0.001329	9.781	0.002297	330.435	0.001138	238.929	0.006360	441.757
189.54	0.001406	10.052	0.002823	330.435	0.001184	248.593	0.007420	444.857
174.18	0.001477	10.270	0.003357	330.435	0.001223	256.762	0.008480	447.166
161.66	0.001542	10.446	0.003896	330.435	0.001256	263.775	0.009540	448.941
151.24	0.001603	10.589	0.004439	330.435	0.001285	269.874	0.010600	450.338
142.41	0.001660	10.706	0.004986	330.435	0.001311	275.235	0.011660	451.461
134.82	0.001715	10.799	0.005536	330.435	0.001333	279.990	0.012720	452.379
128.22	0.001767	10.874	0.006088	330.435	0.001354	284.241	0.013780	453.139
122.44	0.001817	10.931	0.006642	330.435	0.001372	288.068	0.014840	453.776
117.31	0.001865	10.973	0.007198	330.435	0.001388	291.534	0.015900	454.316
112.74	0.001912	11.002	0.007755	330.435	0.001403	294.692	0.016960	454.777
108.64	0.001958	11.018	0.008314	330.435	0.001417	297.583	0.018020	455.174
104.93	0.002002	11.023	0.008873	330.435	0.001430	300.242	0.019080	455.519
101.57	0.002046	11.023	0.009434	330.435	0.001441	302.699	0.020140	455.821
98.50	0.002088	11.023	0.009996	330.435	0.001452	304.975	0.021200	456.085

TABLE 2.4
M-Φ for RC Section 350×500 ($p_t = 1.5\%$, $p_c = 1.2\%$, $f_{ck} = 25$ N/mm^2, $f_y = 380$ N/mm^2)

x_c (mm)	$\varepsilon_{c,max}$	$\sigma_{c,max}$ (kN/m^2)	ε_{st}	σ_{st} (kN/m^2)	ε_{sc}	σ_{sc} (kN/m^2)	Φ (rad/m)	M (kN-m)
195.46	0.000002	0.022	0.000003	0.577	0.000002	0.347	0.000010	0.600
197.34	0.000257	2.647	0.000354	74.436	0.000218	45.684	0.001300	77.401
199.35	0.000518	4.973	0.000704	147.775	0.000440	92.465	0.002600	153.515
201.48	0.000786	6.960	0.001047	219.918	0.000669	140.442	0.003900	228.230
203.75	0.001059	8.586	0.001385	290.749	0.000903	189.731	0.005200	301.417
194.12	0.001262	9.522	0.001793	330.435	0.001067	224.020	0.006500	344.196
173.58	0.001354	9.873	0.002312	330.435	0.001120	235.192	0.007800	347.491
157.90	0.001437	10.150	0.002840	330.435	0.001164	244.417	0.009100	349.840
145.47	0.001513	10.370	0.003375	330.435	0.001201	252.194	0.010400	351.584
135.35	0.001584	10.546	0.003915	330.435	0.001233	258.855	0.011700	352.921
126.94	0.001650	10.686	0.004460	330.435	0.001260	264.638	0.013000	353.970
119.82	0.001713	10.797	0.005008	330.435	0.001284	269.715	0.014300	354.812
113.70	0.001774	10.882	0.005558	330.435	0.001306	274.214	0.015600	355.498
108.40	0.001832	10.946	0.006111	330.435	0.001325	278.234	0.016900	356.065
103.75	0.001888	10.989	0.006666	330.435	0.001342	281.854	0.018200	356.538
99.63	0.001943	11.014	0.007222	330.435	0.001358	285.135	0.019500	356.939
95.96	0.001996	11.023	0.007780	330.435	0.001372	288.127	0.020800	357.280
92.67	0.002048	11.023	0.008339	330.435	0.001385	290.871	0.022100	357.573
89.71	0.002099	11.023	0.008899	330.435	0.001397	293.396	0.023400	357.827
87.01	0.002149	11.023	0.009460	330.435	0.001408	295.728	0.024700	358.048
84.56	0.002199	11.023	0.010021	330.435	0.001419	297.889	0.026000	358.242

TABLE 2.5
M-Φ for RC Section 350×600 ($p_t = 1.5\%$, $p_c = 1.2\%$, $f_{ck} = 25$ N/mm^2, $f_y = 380$ N/mm^2)

x_c (mm)	$\varepsilon_{c,max}$	$\sigma_{c,max}$ (kN/m^2)	ε_{st}	σ_{st} (kN/m^2)	ε_{sc}	σ_{sc} (kN/m^2)	Φ (rad/m)	M (kN-m)
235.47	0.000002	0.026	0.000003	0.703	0.000002	0.431	0.000010	1.083
237.70	0.000253	2.614	0.000354	74.318	0.000221	46.453	0.001065	114.523
240.08	0.000511	4.916	0.000703	147.573	0.000447	93.969	0.002130	227.229
242.60	0.000775	6.889	0.001046	219.671	0.000679	142.642	0.003195	337.963
245.27	0.001045	8.509	0.001383	290.502	0.000917	192.582	0.004260	446.545
233.50	0.001243	9.446	0.001792	330.435	0.001084	227.569	0.005325	510.468
208.28	0.001331	9.790	0.002311	330.435	0.001139	239.227	0.006390	515.502
188.98	0.001409	10.060	0.002841	330.435	0.001185	248.885	0.007455	519.101
173.67	0.001480	10.277	0.003377	330.435	0.001224	257.046	0.008520	521.782
161.18	0.001545	10.453	0.003919	330.435	0.001257	264.051	0.009585	523.841

(Continued)

TABLE 2.5 (CONTINUED)
M-Φ for RC Section 350×600 ($p_t = 1.5\%$, $p_c = 1.2\%$, $f_{ck} = 25$ N/mm², $f_y = 380$ N/mm²)

x_c (mm)	$\varepsilon_{c,max}$	$\sigma_{c,max}$ (kN/m²)	ε_{st}	σ_{st} (kN/m²)	ε_{sc}	σ_{sc} (kN/m²)	Φ (rad/m)	M (kN-m)
150.79	0.001606	10.595	0.004465	330.435	0.001286	270.143	0.010650	525.463
141.98	0.001663	10.711	0.005014	330.435	0.001312	275.496	0.011715	526.766
134.42	0.001718	10.804	0.005567	330.435	0.001334	280.243	0.012780	527.830
127.85	0.001770	10.878	0.006122	330.435	0.001355	284.487	0.013845	528.711
122.08	0.001820	10.934	0.006679	330.435	0.001373	288.307	0.014910	529.450
116.97	0.001869	10.976	0.007237	330.435	0.001389	291.767	0.015975	530.076
112.42	0.001916	11.004	0.007797	330.435	0.001404	294.919	0.017040	530.611
108.33	0.001961	11.019	0.008359	330.435	0.001418	297.804	0.018105	531.071
104.64	0.002006	11.023	0.008921	330.435	0.001431	300.459	0.019170	531.471
101.28	0.002049	11.023	0.009484	330.435	0.001442	302.910	0.020235	531.820
98.23	0.002092	11.023	0.010049	330.435	0.001453	305.181	0.021300	532.127

TABLE 2.6
M-Φ for RC Section 350×700 ($p_t = 1.5\%$, $p_c = 1.2\%$, $f_{ck} = 25$ N/mm², $f_y = 380$ N/mm²)

x_c (mm)	$\varepsilon_{c,max}$	$\sigma_{c,max}$ (kN/m²)	ε_{st}	σ_{st} (kN/m²)	ε_{sc}	σ_{sc} (kN/m²)	Φ (rad/m)	M (kN-m)
275.49	0.000003	0.030	0.000004	0.828	0.000002	0.516	0.000010	1.775
278.06	0.000250	2.586	0.000353	74.077	0.000223	46.883	0.000900	158.569
280.80	0.000505	4.868	0.000701	147.119	0.000451	94.801	0.001800	314.709
283.69	0.000766	6.827	0.001043	219.036	0.000685	143.844	0.002700	468.218
286.77	0.001032	8.443	0.001380	289.722	0.000924	194.118	0.003600	618.860
273.27	0.001230	9.388	0.001785	330.435	0.001095	229.889	0.004500	709.364
243.30	0.001314	9.726	0.002304	330.435	0.001152	241.888	0.005400	716.525
220.36	0.001388	9.992	0.002833	330.435	0.001199	251.848	0.006300	721.656
202.14	0.001455	10.206	0.003369	330.435	0.001239	260.280	0.007200	725.486
187.28	0.001517	10.380	0.003910	330.435	0.001274	267.528	0.008100	728.435
174.89	0.001574	10.523	0.004456	330.435	0.001304	273.838	0.009000	730.760
164.39	0.001627	10.641	0.005006	330.435	0.001330	279.389	0.009900	732.632
155.36	0.001678	10.737	0.005558	330.435	0.001354	284.315	0.010800	734.162
147.51	0.001726	10.816	0.006113	330.435	0.001375	288.720	0.011700	735.432
140.61	0.001772	10.880	0.006670	330.435	0.001394	292.686	0.012600	736.498
134.51	0.001816	10.930	0.007229	330.435	0.001411	296.278	0.013500	737.402
129.06	0.001858	10.968	0.007790	330.435	0.001426	299.549	0.014400	738.176
124.16	0.001900	10.996	0.008351	330.435	0.001441	302.542	0.015300	738.843
119.74	0.001940	11.013	0.008914	330.435	0.001454	305.293	0.016200	739.424
115.72	0.001979	11.022	0.009478	330.435	0.001466	307.832	0.017100	739.931
112.06	0.002017	11.023	0.010043	330.435	0.001477	310.183	0.018000	740.378

TABLE 2.7
M-Φ for RC Section 300 × 450 (p_t = 1.5%, p_c = 1.2%, f_{ck} = 25 N/mm², f_y = 415 N/mm²)

x_c	$\varepsilon_{c,max}$	$\sigma_{c,max}$	ε_{st}	σ_{st}	ε_{sc}	σ_{sc}	Φ	M
(mm)		(kN/m²)		(kN/m²)		(kN/m²)	(rad/m)	(kN-m)
175.45	0.000002	0.019	0.000002	0.514	0.000001	0.305	0.000010	0.364
177.18	0.000262	2.701	0.000359	75.468	0.000218	45.744	0.001480	53.443
179.03	0.000530	5.068	0.000713	149.785	0.000441	92.639	0.002960	105.955
181.00	0.000804	7.079	0.001061	222.845	0.000670	140.791	0.004440	157.455
183.09	0.001084	8.711	0.001402	294.522	0.000906	190.326	0.005920	207.844
183.99	0.001362	9.900	0.001746	360.870	0.001140	239.300	0.007400	254.523
164.88	0.001464	10.232	0.002266	360.870	0.001198	251.517	0.008880	257.127
150.26	0.001557	10.482	0.002794	360.870	0.001246	261.645	0.010360	258.986
138.68	0.001642	10.670	0.003331	360.870	0.001287	270.217	0.011840	260.369
129.24	0.001721	10.810	0.003873	360.870	0.001322	277.591	0.013320	261.429
121.38	0.001796	10.909	0.004420	360.870	0.001352	284.018	0.014800	262.262
114.73	0.001868	10.975	0.004970	360.870	0.001379	289.684	0.016280	262.930
109.02	0.001936	11.012	0.005523	360.870	0.001403	294.729	0.017760	263.474
104.07	0.002002	11.023	0.006079	360.870	0.001425	299.259	0.019240	263.923
99.72	0.002066	11.023	0.006636	360.870	0.001445	303.357	0.020720	264.299
95.87	0.002128	11.023	0.007196	360.870	0.001462	307.084	0.022200	264.616
92.44	0.002189	11.023	0.007757	360.870	0.001479	310.486	0.023680	264.886
89.35	0.002248	11.023	0.008319	360.870	0.001493	313.606	0.025160	265.120
86.57	0.002306	11.023	0.008883	360.870	0.001507	316.476	0.026640	265.322
84.04	0.002363	11.023	0.009447	360.870	0.001520	319.126	0.028120	265.499
81.73	0.002419	11.023	0.010013	360.870	0.001531	321.580	0.029600	265.655

TABLE 2.8
M-Φ for RC Section 300 × 500 (p_t = 1.5%, p_c = 1.2%, f_{ck} = 25 N/mm², f_y = 415 N/mm²)

x_c	$\varepsilon_{c,max}$	$\sigma_{c,max}$	ε_{st}	σ_{st}	ε_{sc}	σ_{sc}	Φ	M
(mm)		(kN/m²)		(kN/m²)		(kN/m²)	(rad/m)	(kN-m)
195.46	0.000002	0.022	0.000003	0.577	0.000002	0.347	0.000010	0.514
197.36	0.000260	2.675	0.000359	75.288	0.000220	46.218	0.001315	67.103
199.40	0.000524	5.023	0.000712	149.453	0.000446	93.559	0.002630	133.076
201.56	0.000795	7.023	0.001059	222.392	0.000677	142.126	0.003945	197.820
203.85	0.001072	8.652	0.001400	293.984	0.000914	192.040	0.005260	261.219
205.03	0.001348	9.852	0.001742	360.870	0.001151	241.670	0.006575	320.485
183.43	0.001447	10.182	0.002261	360.870	0.001211	254.225	0.007890	323.833
166.91	0.001536	10.431	0.002790	360.870	0.001260	264.652	0.009205	326.228
153.80	0.001618	10.621	0.003326	360.870	0.001302	273.489	0.010520	328.013
143.10	0.001694	10.765	0.003869	360.870	0.001339	281.096	0.011835	329.385

(Continued)

TABLE 2.8 (CONTINUED)
M-Φ for RC Section 300 × 500 ($p_t = 1.5\%$, $p_c = 1.2\%$, $f_{ck} = 25$ N/mm², $f_y = 415$ N/mm²)

x_c	$\varepsilon_{c,max}$	$\sigma_{c,max}$	ε_{st}	σ_{st}	ε_{sc}	σ_{sc}	Φ	M
(mm)		(kN/m²)		(kN/m²)		(kN/m²)	(rad/m)	(kN-m)
134.19	0.001765	10.871	0.004416	360.870	0.001370	287.731	0.013150	330.466
126.65	0.001832	10.946	0.004967	360.870	0.001398	293.580	0.014465	331.333
120.16	0.001896	10.994	0.005520	360.870	0.001423	298.786	0.015780	332.042
114.53	0.001958	11.019	0.006077	360.870	0.001445	303.458	0.017095	332.628
109.58	0.002017	11.023	0.006635	360.870	0.001465	307.681	0.018410	333.119
105.21	0.002075	11.023	0.007196	360.870	0.001483	311.520	0.019725	333.534
101.30	0.002131	11.023	0.007757	360.870	0.001500	315.027	0.021040	333.889
97.79	0.002186	11.023	0.008321	360.870	0.001515	318.243	0.022355	334.194
94.62	0.002240	11.023	0.008885	360.870	0.001530	321.202	0.023670	334.460
91.74	0.002292	11.023	0.009451	360.870	0.001543	323.934	0.024985	334.693
89.11	0.002344	11.023	0.010017	360.870	0.001555	326.465	0.026300	334.897

TABLE 2.9
M-Φ for RC Section 300 × 600 ($p_t = 1.5\%$, $p_c = 1.2\%$, $f_{ck} = 25$ N/mm², $f_y = 415$ N/mm²)

x_c	$\varepsilon_{c,max}$	$\sigma_{c,max}$	ε_{st}	σ_{st}	ε_{sc}	σ_{sc}	Φ	M
(mm)		(kN/m²)		(kN/m²)		(kN/m²)	(rad/m)	(kN-m)
235.47	0.000002	0.026	0.000003	0.703	0.000002	0.431	0.000010	0.929
237.73	0.000256	2.637	0.000357	75.011	0.000223	46.894	0.001075	99.077
240.13	0.000516	4.957	0.000709	148.938	0.000452	94.872	0.002150	196.566
242.67	0.000783	6.939	0.001056	221.684	0.000686	144.031	0.003225	292.330
245.38	0.001055	8.563	0.001396	293.136	0.000926	194.484	0.004300	386.210
247.15	0.001328	9.780	0.001735	360.870	0.001167	245.104	0.005375	475.224
220.59	0.001423	10.105	0.002254	360.870	0.001229	258.153	0.006450	480.345
200.24	0.001507	10.353	0.002782	360.870	0.001281	269.018	0.007525	484.021
184.07	0.001583	10.544	0.003319	360.870	0.001325	278.243	0.008600	486.770
170.86	0.001653	10.692	0.003862	360.870	0.001363	286.195	0.009675	488.888
159.85	0.001718	10.805	0.004409	360.870	0.001396	293.136	0.010750	490.561
150.51	0.001780	10.890	0.004960	360.870	0.001425	299.259	0.011825	491.908
142.48	0.001838	10.951	0.005515	360.870	0.001451	304.708	0.012900	493.011
135.49	0.001894	10.992	0.006072	360.870	0.001474	309.594	0.013975	493.926
129.35	0.001947	11.016	0.006632	360.870	0.001495	314.007	0.015050	494.694
123.91	0.001998	11.023	0.007193	360.870	0.001514	318.016	0.016125	495.346
119.06	0.002048	11.023	0.007756	360.870	0.001532	321.677	0.017200	495.903
114.69	0.002096	11.023	0.008321	360.870	0.001548	325.035	0.018275	496.385
110.75	0.002143	11.023	0.008886	360.870	0.001563	328.126	0.019350	496.803
107.17	0.002189	11.023	0.009453	360.870	0.001576	330.981	0.020425	497.170
103.89	0.002234	11.023	0.010021	360.870	0.001589	333.625	0.021500	497.493

TABLE 2.10
M-Φ for RC Section 300 × 600 (p_t = 2.0%, p_c = 1.5%, f_{ck} = 25 N/mm², f_y = 415 N/mm²)

x_c (mm)	$\varepsilon_{c,max}$	$\sigma_{c,max}$ (kN/m²)	ε_{st}	σ_{st} (kN/m²)	ε_{sc}	σ_{sc} (kN/m²)	Φ (rad/m)	M (kN-m)
253.84	0.000003	0.028	0.000003	0.664	0.000002	0.470	0.000010	1.168
256.29	0.000287	2.937	0.000351	73.784	0.000253	53.224	0.001120	129.721
258.91	0.000580	5.466	0.000697	146.338	0.000513	107.678	0.002240	257.112
261.68	0.000879	7.562	0.001036	217.548	0.000778	163.476	0.003360	381.970
264.64	0.001186	9.196	0.001368	287.283	0.001051	220.749	0.004480	504.059
267.80	0.001500	10.334	0.001692	355.385	0.001332	279.655	0.005600	623.099
240.89	0.001619	10.623	0.002212	360.870	0.001417	297.605	0.006720	639.282
217.95	0.001709	10.790	0.002760	360.870	0.001474	309.448	0.007840	644.483
199.76	0.001790	10.902	0.003317	360.870	0.001521	319.425	0.008960	648.340
184.94	0.001864	10.973	0.003881	360.870	0.001562	327.968	0.010080	651.290
172.59	0.001933	11.011	0.004451	360.870	0.001597	335.378	0.011200	653.603
162.14	0.001998	11.023	0.005025	360.870	0.001628	341.878	0.012320	655.454
153.17	0.002059	11.023	0.005602	360.870	0.001655	347.633	0.013440	656.959
145.37	0.002117	11.023	0.006183	360.870	0.001680	352.765	0.014560	658.202
138.53	0.002172	11.023	0.006765	360.870	0.001702	357.370	0.015680	659.242
132.98	0.002234	11.023	0.007342	360.870	0.001730	360.870	0.016800	660.015
130.50	0.002339	11.023	0.007876	360.870	0.001801	360.870	0.017920	660.173
128.31	0.002443	11.023	0.008410	360.870	0.001872	360.870	0.019040	660.303
126.37	0.002548	11.023	0.008944	360.870	0.001943	360.870	0.020160	660.413
124.63	0.002652	11.023	0.009478	360.870	0.002014	360.870	0.021280	660.505
123.06	0.002757	11.023	0.010011	360.870	0.002085	360.870	0.022400	660.584

TABLE 2.11
M-Φ for RC Section 300 × 600 (p_t = 2.5%, p_c = 2.0%, f_{ck} = 25 N/mm², f_y = 415 N/mm²)

x_c (mm)	$\varepsilon_{c,max}$	$\sigma_{c,max}$ (kN/m²)	ε_{st}	σ_{st} (kN/m²)	ε_{sc}	σ_{sc} (kN/m²)	Φ (rad/m)	M (kN-m)
260.56	0.000003	0.029	0.000003	0.650	0.000002	0.484	0.000010	1.436
262.80	0.000294	3.006	0.000344	72.254	0.000261	54.754	0.001120	159.655
265.17	0.000594	5.575	0.000683	143.391	0.000527	110.625	0.002240	316.747
267.67	0.000899	7.685	0.001016	213.321	0.000799	167.703	0.003360	471.078
270.32	0.001211	9.308	0.001343	281.939	0.001077	226.093	0.004480	622.425
273.13	0.001530	10.413	0.001662	349.123	0.001362	285.917	0.005600	770.527
247.63	0.001664	10.712	0.002166	360.870	0.001462	307.115	0.006720	803.297
222.91	0.001748	10.848	0.002721	360.870	0.001512	317.603	0.007840	809.358
203.41	0.001823	10.937	0.003285	360.870	0.001554	326.291	0.008960	813.776
187.61	0.001891	10.991	0.003855	360.870	0.001589	333.620	0.010080	817.106

(Continued)

TABLE 2.11 (CONTINUED)

M-Φ for RC Section 300 × 600 ($p_t = 2.5\%$, $p_c = 2.0\%$, $f_{ck} = 25$ N/mm², $f_y = 415$ N/mm²)

x_c (mm)	$\varepsilon_{c,max}$	$\sigma_{c,max}$ (kN/m²)	ε_{st}	σ_{st} (kN/m²)	ε_{sc}	σ_{sc} (kN/m²)	Φ (rad/m)	M (kN-m)
174.51	0.001955	11.018	0.004429	360.870	0.001619	339.895	0.011200	819.683
163.48	0.002014	11.023	0.005008	360.870	0.001644	345.334	0.012320	821.719
154.04	0.002070	11.023	0.005590	360.870	0.001667	350.097	0.013440	823.359
145.88	0.002124	11.023	0.006175	360.870	0.001687	354.304	0.014560	824.699
138.74	0.002175	11.023	0.006762	360.870	0.001705	358.046	0.015680	825.810
132.98	0.002234	11.023	0.007342	360.870	0.001730	360.870	0.016800	826.628
130.50	0.002339	11.023	0.007876	360.870	0.001801	360.870	0.017920	826.786
128.31	0.002443	11.023	0.008410	360.870	0.001872	360.870	0.019040	826.917
126.37	0.002548	11.023	0.008944	360.870	0.001943	360.870	0.020160	827.026
124.63	0.002652	11.023	0.009478	360.870	0.002014	360.870	0.021280	827.119
123.06	0.002757	11.023	0.010011	360.870	0.002085	360.870	0.022400	827.198

TABLE 2.12

M-Φ for RC Section 350 × 500 ($p_t = 1.5\%$, $p_c = 1.2\%$, $f_{ck} = 25$ N/mm², $f_y = 415$ N/mm²)

x_c (mm)	$\varepsilon_{c,max}$	$\sigma_{c,max}$ (kN/m²)	ε_{st}	σ_{st} (kN/m²)	ε_{sc}	σ_{sc} (kN/m²)	Φ (rad/m)	M (kN-m)
195.46	0.000002	0.022	0.000003	0.577	0.000002	0.347	0.000010	0.600
197.36	0.000259	2.671	0.000358	75.146	0.000220	46.129	0.001313	78.139
199.39	0.000523	5.015	0.000710	149.174	0.000445	93.376	0.002625	154.965
201.54	0.000794	7.012	0.001057	221.980	0.000675	141.845	0.003938	230.364
203.84	0.001070	8.641	0.001397	293.445	0.000913	191.655	0.005250	304.199
205.27	0.001347	9.849	0.001737	360.870	0.001150	241.538	0.006563	373.854
183.65	0.001446	10.178	0.002255	360.870	0.001210	254.095	0.007875	377.767
167.10	0.001535	10.428	0.002783	360.870	0.001260	264.525	0.009188	380.568
153.97	0.001617	10.619	0.003318	360.870	0.001302	273.364	0.010500	382.654
143.27	0.001692	10.763	0.003860	360.870	0.001338	280.975	0.011813	384.259
134.35	0.001763	10.869	0.004405	360.870	0.001370	287.612	0.013125	385.522
126.79	0.001831	10.944	0.004955	360.870	0.001397	293.465	0.014438	386.537
120.30	0.001895	10.993	0.005508	360.870	0.001422	298.674	0.015750	387.365
114.66	0.001956	11.018	0.006063	360.870	0.001445	303.348	0.017063	388.051
109.71	0.002016	11.023	0.006620	360.870	0.001465	307.573	0.018375	388.625
105.32	0.002074	11.023	0.007180	360.870	0.001483	311.416	0.019688	389.110
101.41	0.002130	11.023	0.007740	360.870	0.001500	314.925	0.021000	389.525
97.90	0.002184	11.023	0.008303	360.870	0.001515	318.143	0.022313	389.883
94.72	0.002238	11.023	0.008866	360.870	0.001529	321.105	0.023625	390.194
91.84	0.002290	11.023	0.009430	360.870	0.001542	323.839	0.024938	390.466
89.21	0.002342	11.023	0.009996	360.870	0.001554	326.372	0.026250	390.705

TABLE 2.13

M-Φ for RC Section 350 × 600 (p$_t$ = 1.5%, p$_c$ = 1.2%, f$_{ck}$ = 25 N/mm², f$_y$ = 415 N/mm²)

x$_c$	ε$_{c,max}$	σ$_{c,max}$	ε$_{st}$	σ$_{st}$	ε$_{sc}$	σ$_{sc}$	Φ	M
(mm)		(kN/m²)		(kN/m²)		(kN/m²)	(rad/m)	(kN-m)
235.47	0.000002	0.026	0.000003	0.703	0.000002	0.431	0.000010	1.083
237.73	0.000256	2.637	0.000357	75.011	0.000223	46.894	0.001075	115.590
240.13	0.000516	4.957	0.000709	148.938	0.000452	94.872	0.002150	229.327
242.67	0.000783	6.939	0.001056	221.684	0.000686	144.031	0.003225	341.052
245.38	0.001055	8.563	0.001396	293.136	0.000926	194.484	0.004300	450.579
247.15	0.001328	9.780	0.001735	360.870	0.001167	245.104	0.005375	554.428
220.59	0.001423	10.105	0.002254	360.870	0.001229	258.153	0.006450	560.402
200.24	0.001507	10.353	0.002782	360.870	0.001281	269.018	0.007525	564.692
184.07	0.001583	10.544	0.003319	360.870	0.001325	278.243	0.008600	567.898
170.86	0.001653	10.692	0.003862	360.870	0.001363	286.195	0.009675	570.370
159.85	0.001718	10.805	0.004409	360.870	0.001396	293.136	0.010750	572.321
150.51	0.001780	10.890	0.004960	360.870	0.001425	299.259	0.011825	573.893
142.48	0.001838	10.951	0.005515	360.870	0.001451	304.708	0.012900	575.180
135.49	0.001894	10.992	0.006072	360.870	0.001474	309.594	0.013975	576.247
129.35	0.001947	11.016	0.006632	360.870	0.001495	314.007	0.015050	577.143
123.91	0.001998	11.023	0.007193	360.870	0.001514	318.016	0.016125	577.903
119.06	0.002048	11.023	0.007756	360.870	0.001532	321.677	0.017200	578.554
114.69	0.002096	11.023	0.008321	360.870	0.001548	325.035	0.018275	579.115
110.75	0.002143	11.023	0.008886	360.870	0.001563	328.126	0.019350	579.604
107.17	0.002189	11.023	0.009453	360.870	0.001576	330.981	0.020425	580.031
103.89	0.002234	11.023	0.010021	360.870	0.001589	333.625	0.021500	580.408

TABLE 2.14

M-Φ for RC Section 350 × 700 (p$_t$ = 1.5%, p$_c$ = 1.2%, f$_{ck}$ = 25 N/mm², f$_y$ = 415 N/mm²)

x$_c$	ε$_{c,max}$	σ$_{c,max}$	ε$_{st}$	σ$_{st}$	ε$_{sc}$	σ$_{sc}$	Φ	M
(mm)		(kN/m²)		(kN/m²)		(kN/m²)	(rad/m)	(kN-m)
275.49	0.000003	0.030	0.000004	0.828	0.000002	0.516	0.000010	1.775
278.09	0.000253	2.613	0.000357	74.894	0.000226	47.410	0.000910	160.317
280.86	0.000511	4.915	0.000708	148.730	0.000457	95.878	0.001820	318.150
283.79	0.000775	6.886	0.001054	221.412	0.000693	145.500	0.002730	473.287
286.91	0.001044	8.507	0.001394	292.833	0.000935	196.383	0.003640	625.484
289.09	0.001315	9.732	0.001733	360.870	0.001179	247.558	0.004550	770.508
257.59	0.001406	10.052	0.002252	360.870	0.001243	260.950	0.005460	778.985
233.42	0.001487	10.298	0.002781	360.870	0.001296	272.118	0.006370	785.085
214.20	0.001559	10.488	0.003318	360.870	0.001341	281.611	0.007280	789.656
198.50	0.001626	10.637	0.003862	360.870	0.001380	289.803	0.008190	793.185

(Continued)

TABLE 2.14 (CONTINUED)

M-Φ for RC Section 350 × 700 (p_t = 1.5%, p_c = 1.2%, f_{ck} = 25 N/mm², f_y = 415 N/mm²)

x_c	$\varepsilon_{c,max}$	$\sigma_{c,max}$	ε_{st}	σ_{st}	ε_{sc}	σ_{sc}	Φ	M
(mm)		(kN/m²)		(kN/m²)		(kN/m²)	(rad/m)	(kN-m)
185.39	0.001687	10.754	0.004410	360.870	0.001414	296.958	0.009100	795.977
174.27	0.001744	10.843	0.004962	360.870	0.001444	303.272	0.010010	798.229
164.70	0.001799	10.912	0.005518	360.870	0.001471	308.891	0.010920	800.075
156.37	0.001850	10.961	0.006076	360.870	0.001495	313.931	0.011830	801.610
149.04	0.001899	10.995	0.006637	360.870	0.001517	318.479	0.012740	802.900
142.54	0.001946	11.015	0.007200	360.870	0.001536	322.610	0.013650	803.996
136.74	0.001991	11.023	0.007764	360.870	0.001554	326.379	0.014560	804.936
131.53	0.002035	11.023	0.008330	360.870	0.001571	329.837	0.015470	805.747
126.81	0.002077	11.023	0.008897	360.870	0.001586	333.019	0.016380	806.454
122.53	0.002119	11.023	0.009466	360.870	0.001600	335.958	0.017290	807.073
118.61	0.002159	11.023	0.010035	360.870	0.001613	338.681	0.018200	807.619

2.9 CONCLUSIONS

In this chapter, a new analytical procedure for estimating curvature ductility of RC sections is proposed. The purpose is to estimate the moment-curvature relationship under service loads, in a simpler closed-form manner. Analytical expressions for moment-curvature relationship of RC sections, accounting for nonlinear characteristics of constitutive materials according to Eurocode, are proposed in elastic and elastic-plastic ranges as well. Percentage of tension reinforcement influences curvature ductility to a large extent. There exists at least one critical value of percentage of both tensile and compression reinforcements that reduces the curvature ductility to the minimum. The proposed analytical expressions are capable of tracing this critical value, so that it can be avoided for a successful design of the section. Tensile reinforcement, closer to $p_{t,bal}$, will result in more curvature ductility since there is a marginal reduction seen due to the kink in the curve for (lesser) values closer to $p_{t,bal}$. Maximum curvature ductility is obtained for compression reinforcement equal to $p_{c,bal}$ when the section is subjected to axial compressive forces; for tensile axial forces, the percentage of compression steel, the same as that of tension steel ($p_c = p_t$), gives the maximum curvature ductility.

With regard to their close agreement with the analytical procedure, proposed expressions for moment-curvature estimate are thus qualified for use in design and in structural assessments as well. Avoiding tedious hand-calculations and approximations required in conventional iterative design procedures, the proposed method eliminates the possibility of potentially unsafe design. In the absence of enough experimental evidence to be more conclusive on the topic, the proposed closed-form solutions for the unknown curvature ductility ratios are confident of giving a reliable and safe estimate of the said parameter. With due consideration to the increasing necessity of structural assessment of existing buildings under seismic loads, the proposed expressions of moment-curvature relationship shall become an integral input while employing nonlinear static procedures.

2.10   SPREADSHEET PROGRAM

The presented analytical expressions for moment-curvature relationship in closed form are very useful for researchers. However, to facilitate the ready application of the developed procedure, a simple spreadsheet program used to estimate the moment-curvature relationship is presented; this should encourage the structural designers to use it instantly and with confidence. The spreadsheet is available in the CD content, which can be freely downloaded from the following Web site: http://www.crcpress .com/e_products/downloads/download.asp?cat_no=K10453. Table 2.15 shows the values of the points traced along the M-Φ curve, obtained numerically, for no axial force case.

2.10.1   Step-by-Step Procedure to Use the Spreadsheet Program Given on the Web Site

Table 2.15 shows the demonstrated example case, which is explained in this section.
    First, to predict the moment-curvature relationship in elastic range, the steps are as follows:

1. An arbitrary value is assumed for the limit elastic curvature; assign any value to cell B21; for example, 0.005.
2. Fix axial force to the desired value; depth of neutral axis is determined. Click cell A21; Go to Tools in the menu; then select Goal Seek; Set cell: A21; To value: axial force; for example, axial force is set to zero; By changing cell: click iteration, select C21 (neutral axis position); click iteration. You will find a remark: Goal Seeking with Cell A21 found solution. If target value and current value are the same, then the solution is determined; press OK. We get x_c as 0.172. Observe the values of cells D21 = 0.00086; E21 = 0.000708; F21 = 0.00149. These values correspond to $\varepsilon_{c,max}, \varepsilon_{sc}, \varepsilon_{st}$, respectively. These values of the strain should be less than $\varepsilon_{c0} = (\text{cell I11}); \varepsilon_{s0} = (\text{cell M11})$. In this case, $\varepsilon_{s0} = 0.00172$. Therefore, we increase the curvature to 0.00578 to get x_c as 0.173 m for $\varepsilon_{st} = \varepsilon_{s0}$. Thus, the limit elastic curvature is determined as 0.00578 rad/m, and the corresponding moment is 232.62 kN-m.
3. Fixing this value as the limit elastic curvature and subdividing it equally, moment-curvature values for the first five rows are now obtained as follows: For example, consider the first row, select Cell A16; Go to Tools; select Goal Seek; Set Cell A16; To value: Axial force (in this case it is zero); By changing Cell: Click iteration; select C16; press OK. You will find a remark: Goal Seeking with Cell A21 found solution. If target value

TABLE 2.15

M-Φ for RC Section 300 × 500 for No Axial Force (p_t = 1.08%, p_c = 1.08%, R_{ck} = 30 N/mm², f_y = 415 N/mm²)

P (kN)	Φ (rad/m)	x_c (m)	$\varepsilon_{c,max}$	ε_{sc}	ε_{st}	$\sigma_{c,max}$ (kN/m²)	σ_{sc} (kN/m²)	σ_{st} (kN/m²)	q (m)	M (kN-m)
0.00	0.000010	0.165	0.00000	0.000001	0.00000	22	284	640	0.00	0.41
0.00	0.001166	0.167	0.00019	0.000159	0.00035	2444	33437	74301	0.00	48.07
0.00	0.002322	0.168	0.00039	0.000320	0.00070	4657	67283	147269	0.00	95.20
0.00	0.003478	0.169	0.00059	0.000485	0.00105	6648	101874	219493	0.00	141.76
0.00	0.004634	0.171	0.00079	0.000654	0.00139	8408	137269	290913	0.00	187.72
0.00	0.005780	0.173	0.00100	0.000825	0.00172	9909	173216	360856	0.00	232.62
0.00	0.007080	0.153	0.00108	0.000872	0.00224	10457	183156	360870	0.00	234.86
0.00	0.008379	0.139	0.00116	0.000911	0.00278	10906	191230	360870	0.00	236.41
0.00	0.009679	0.127	0.00123	0.000943	0.00332	11283	197947	360870	0.00	237.55
0.00	0.010979	0.118	0.00130	0.000970	0.00386	11603	203635	360870	0.00	238.40
0.00	0.012279	0.111	0.00136	0.000993	0.00441	11879	208523	360870	0.00	239.07
0.00	0.013578	0.105	0.00142	0.001013	0.00496	12118	212773	360870	0.00	239.61
0.00	0.014878	0.099	0.00148	0.001031	0.00552	12325	216505	360870	0.00	240.04
0.00	0.016178	0.095	0.00153	0.001047	0.00607	12504	219811	360870	0.00	240.40
0.00	0.017478	0.091	0.00159	0.001061	0.00663	12659	222762	360870	0.00	240.69
0.00	0.018777	0.087	0.00164	0.001073	0.00719	12792	225415	360870	0.00	240.95
0.00	0.020077	0.084	0.00169	0.001085	0.00775	12904	227814	360870	0.00	241.16
0.00	0.021377	0.081	0.00174	0.001095	0.00831	12999	229997	360870	0.00	241.34
0.00	0.022677	0.079	0.00179	0.001105	0.00887	13075	231994	360870	0.00	241.50
0.00	0.023976	0.076	0.00183	0.001113	0.00944	13136	233829	360870	0.00	241.64
0.00	0.025276	0.074	0.00188	0.001122	0.01000	13180	235525	360870	0.00	241.77

and current value are the same; then the solution is determined; press OK. Repeat the same procedure for the next four rows and obtain the points of M-Φ curve in elastic range.

Second, for estimating the values in elastic-plastic range, the following steps are adopted:

1. Assign any value to cell B36; for example, 0.025.
2. Fix axial force to the desired value; depth of neutral axis is determined. Click A36; Go to Tools in the menu; then select Goal seek; Set cell: A36; To value: axial force; for example, axial force is set to zero; By changing cell: click iteration; select C36 (neutral axis position); click iteration; You will find a remark: Goal Seeking with Cell A36 found solution; If target value and current value are the same; then the solution is determined; press OK. We get x_c as 0.075. Observe the values of cells D36 = 0.00187; E36 = 0.00112; F36 = 0.00988. These values correspond to $\varepsilon_{c,max}, \varepsilon_{sc}, \varepsilon_{st}$, respectively. These values of the strain should be less than $\varepsilon_{cu} = $(cell J11); $\varepsilon_{su} = $(cell N11). In this case, $\varepsilon_{su} = 0.01$. Therefore, we increase the curvature to 0.025276 to get x_c as 0.074 m for $\varepsilon_{st} = \varepsilon_{su}$. Thus, ultimate curvature is determined as 0.025276 rad/m, and the corresponding moment is 241.77 kNm.
3. Fixing this value as the ultimate curvature and subdividing it equally, moment-curvature values for the next 14 rows after limit elastic values (first yellow band) are now obtained as follows: For example, consider row 22; select Cell A22; Go to Tools; select Goal Seek; Set Cell A22; To value: Axial force (in this case it is zero); By changing Cell: Click iteration, select C22; press OK. You will find a remark: Goal Seeking with Cell A22 found solution. If target value and current value are the same, then the solution is determined; press OK. Repeat the same procedure for the next 13 rows and obtain the points of M-Φ curve in elastic-plastic range.

The above example shows that tensile strain in steel reaches limit elastic values and ultimate value first, making the failure as tension failure. However, in some examples, you may also see that concrete reaches its maximum value first, making it as a compression failure.

3 Moment-Rotation Relationship for RC Beams

3.1 SUMMARY

Moment-rotation relationships of RC beams provide an estimate of the beams' ductility, which is a valuable design parameter. The correct estimate of ductility is very important in the context of recent advancements in design approaches like displacement-based design. In this chapter, collapse mechanism and plastic hinge extensions of RC beams in bending, under increasing concentrated design load until collapse, are examined. Moment-rotation relationships in explicit form, in elastic and elastic-plastic ranges, are derived from the proposed bilinear modeling of moment-curvature relationships, presented in Chapter 2. Analytical estimates are verified for equilibrium and compatibility conditions. Ductility ratios of two cases, (1) a fixed beam and (2) a simply supported beam, are presented. The proposed analytical procedure is capable of modeling the moment-rotation relationship, accounting for nonlinear characteristics of the materials, and providing a satisfactory estimate of ductility. They are useful for designing special moment-resisting RC framed structures, in particular, where ductility is an important design parameter. In the technical context in this chapter, *relative rotation* occurring between the extremities of a plastic hinge is termed as *rotation*.

3.2 INTRODUCTION

Recent revisions in design approaches of RC elements include desirable features of ultimate strength and working stress design as well, to ensure satisfactory design. A seismic design procedure that does not account for maximum plastic deformation demands, which a structure is likely to undergo during severe ground motion, could lead to unreliable performance (Amador and Nadyane 2008). With displacement-based design approach becoming more common, it is imperative to ensure conceptual implication of multiple target performance (damage) levels that are expected to be achieved, or at least not exceeded, when the structure is subjected to earthquakes of specified intensity (Priestley, Calvi, and Kowalsky 2007). Gilbert and Smith (2006) showed the significance of strain localization in RC slabs and its adverse effect on ductility. While seismic design philosophy demands energy dissipation/absorption by postelastic deformation for collapse prevention during major earthquakes, the seismic capacity of buildings is highly sensitive to their ductility estimates (Zhang and Der Kiureghian 1993). Owing to the large economic losses derived from recent seismic events, design methodologies based on explicit control of dynamic response of structures emphasizing sufficient ductility of members at local and global levels are being practiced by and large; this should lead to a desired solution for the sustainable development of building stock envisaging more major earthquakes in future (Mahin et al. 2006). Ductility, a measure of energy dissipation by inelastic deformation during major earthquakes, depends mainly on moment-curvature relationship at critical sections, where plastic hinges are expected/imposed to be formed at collapse; it also ensures effective redistribution of moments at these sections, as collapse load is approached (Park and Paulay 1975; Paulay and Priestley 1992). Damage models that quantify severity of repeated plastic cycling through plastic energy dissipation are simple tools that can be used for practical seismic design. In other words, structures should be designed to resist earthquakes in a quantifiable manner, imposed with desired possible damage (Bangash 1989; Ghobarah 2001; SEAOC 1995). Structural performance of a building during an earthquake depends on many parameters such as material properties and hysteretic behavior of members, joints, and the like that are highly uncertain (Rustem 2006). Fan Sau-Cheong and Wang (2002) justified recommendation of reinforced concrete structures to resist seismic loading only if the design is capable of ensuring sufficient ductility.

The literature reviewed critically emphasizes the importance of ductility in RC sections to ensure satisfactory behavior under seismic loads. However, analytical expressions, in a closed form for moment-rotation relationship and ductility of rectangular RC sections (with different tensile and compressive reinforcements), accounting for nonlinear properties of constitutive materials, are relatively absent in the literature. This chapter presents a mathematical development of nonlinear behavior of RC beams based in Eurocode currently in prevalence and derives moment-rotation relationships and ductility from the bilinear modeling of moment-curvature relationships presented in Chapter 2. Theoretical moment-rotation curves for RC beams in bending, under increasing central concentrated design load until collapse, are studied. Ductility ratios of fixed beams and simply supported beams are examined and discussed.

3.3 MATHEMATICAL DEVELOPMENT

Concrete under multiaxial compressive stress state exhibits significant nonlinearity. The fundamental Bernoulli hypothesis of linear strain over the cross-section for both elastic and elastic-plastic responses of the beam, under bending moment combined with axial force, is assumed in the study. Axial force–bending moment yield interaction discussed in Chapter 1 is recalled. For classifying the failure as *tension (or) compression* caused by yielding of steel (or) crushing of concrete, respectively, the percentage of steel or a balanced section is given by

$$P_{t,bal} = P_c + \frac{(3\varepsilon_{cu} - \varepsilon_{c0})\sigma_{c0}}{3(\varepsilon_{cu} + \varepsilon_{su})\sigma_{s0}} - \frac{P_0}{b(D-d)\sigma_{s0}} \tag{3.1}$$

where P_0 is the axial force ($P_0 > 0$, if it is compression). For the known cross-section with fixed percentage of compression reinforcement, Equation 3.1 gives the percentage of reinforcement for the balanced section. It is to be noted that the above equation is the same as Equation 2.51. Moment-curvature relationship, as presented in Chapter 2, is recalled for elastic and elastic-plastic range. It is well known that the presence of axial force influences moment-curvature relationships. It is also essential to know the critical value of axial force up to which the failure remains tensile; only until those critical values, moment-rotation relationships are proposed to be examined in the following section. For the axial force of $P \in [0, P^*]$, collapse is caused by yielding of tensile steel, and for $P > P^*$, collapse is caused by crushing of concrete. The critical value of axial force P^* is given by

$$P^* = b\,(D-d)\left[\frac{(3\varepsilon_{cu} - \varepsilon_{c0})}{3(\varepsilon_{cu} + \varepsilon_{su})}\,\sigma_{c0} - (p_t - p_c)\sigma_{s0}\right] \qquad (3.2)$$

Moment-curvature relationships, in elastic and elastic-plastic ranges are now examined for an RC beam of cross-section 300×450 mm with R_{ck} as 25 N/mm^2 and f_y as 41 N/mm^2. For members of building frames under seismic loads, in particular, design type leading to tension failure is normally used, because of its known advantages. For the beam reinforced with the same percentage of tension and compression steel (4#22Φ), critical axial force computed from Equation 3.2 amounts to 291.51kN. Moment-curvature relationships are examined for axial forces less than this critical value, and relevant curves are shown in Figure 3.1. Both the curves obtained from bilinear approximation and using moment-curvature relationships are presented in the figure. It is seen from the figure that the variations are very small for lesser values of axial force but tend to increase for greater values. Therefore, the developed moment-curvature relationship discussed in Chapter 2 validates the procedure for tensile failure up to axial load level of critical value. Hence it is satisfactory to use bilinear approximated moment-curvature to further investigate moment-rotation relationships and rotation ductility of RC beams; the same is used in the further sections of discussions.

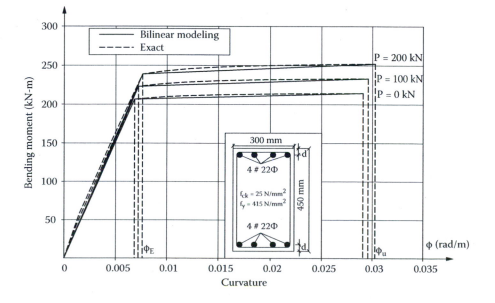

FIGURE 3.1 Moment-curvature for RC section 300×450 for different axial forces.

3.4 ANALYTICAL MOMENT-ROTATION RELATIONSHIPS

The relationships shall be examined in elastic and elastic-plastic range, successively (Nunziante, Gambarotta, and Tralli 2007). General equations of equilibrium are known as

$$\frac{d\,V(z)}{d\,z} = -q(z), \quad \frac{d^2M(z)}{d\,z^2} = -q(z) \tag{3.3}$$

where $V(z)$, $M(z)$, $q(z)$ are shear force, bending moment, and distributed load present in the beam. Assuming the hypothesis of small displacement, compatibility equations are given by

$$\phi(z) = -\frac{d^2\delta(z)}{d\,z^2} \tag{3.4}$$

where, $\delta(z)$ is the transverse displacement function of the beam. Moment-curvature relationships, in elastic and elastic-plastic ranges for monotonically increasing curvature, are known as

$$M(z) = \begin{cases} K_E^\phi\, \phi(z) & \forall \phi \in [0, \phi_E] \\ M_E + K_p^\phi[\phi(z) - \phi_E] & \forall \phi \in [\phi_E, \phi_u] \end{cases} \tag{3.5}$$

where, $K_E^\phi = \frac{M_E}{\phi_E}$ is the curvature-elastic stiffness and $K_p^\phi = \frac{M_u - M_E}{\phi_u - \phi_E}$ is the curvature-hardening modulus. Figure 3.2 shows moment-curvature and moment-relative rotation for the beams considered in the analysis, showing also the elastic stiffness and the hardening modulus for curvature and relative rotation, respectively. General differential equations for the beam can now be written as

$$K_E^\phi\, \frac{d^4\delta_e(z)}{d\,z^4} = q(z) \quad \text{(for elastic range)} \tag{3.6}$$

$$K_p^\phi\, \frac{d^4\delta_p(z)}{d\,z^4} = q(z) \quad \text{(for plastic-hardening range)} \tag{3.7}$$

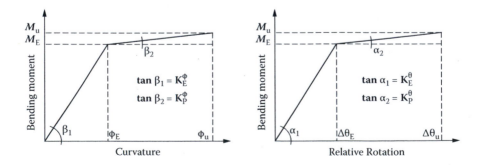

FIGURE 3.2 Elastic stiffness and hardening modulus: (a) relative rotation, (b) curvature.

Further, (1) rotation, (2) curvature, (3) bending moment, and (4) shear in elastic and elastic-plastic range, expressed as the function of displacement, are given by

$$\theta_e(z) = -\frac{d\delta_e(z)}{dz}, \quad \phi_e(z) = -\frac{d^2\delta_e(z)}{dz^2}, \quad M_e(z) = -K_E^\phi \frac{d^2\delta_e(z)}{dz^2}, \quad V_e(z) = -K_E^\phi \frac{d^3\delta_e(z)}{dz^3}$$

(3.8)

$$\theta_p(z) = -\frac{d\delta_p(z)}{dz}, \phi_p(z) = -\frac{d^2\delta_p(z)}{dz^2}, \quad M_p(z) = M_E - K_p^\phi\left[\frac{d^2\delta_p(z)}{dz^2} + \phi_E\right],$$

$$V_p(z) = -K_p^\phi \frac{d^3\delta_p(z)}{dz^3}$$

(3.9)

This proposed modeling of elastic-plastic beam allows to obtain deformations with good accuracy. On the basis of the procedure discussed above and with the help of bilinear approximated moment-curvature, moment-rotation relationships of two cases are now examined.

3.4.1 Fixed Beam under Central Concentrated Load

Figures 3.3 and 3.4 show a fixed beam and simply supported beams under central concentrated load. The beams are examined for increasing design load until collapse and the corresponding moment-rotation relationships, both in elastic and elastic-plastic ranges; axial force is not considered in the analysis (P = 0). In the elastic range, the beam is subdivided in two parts whose lengths lie in the range (0, L/2) and (L/2, L), respectively. Displacement functions of both of the parts in the same reference system (with origin at left support of the beam) are given by

$$\delta_1(z) = S_{01} + S_{11}z + S_{21}z^2 + S_{31}z^3 \qquad\qquad \forall z \in [0, L/2]$$
$$\delta_2(z) = S_{02} + S_{12}(z - L/2) + S_{22}(z - L/2)^2 + S_{32}(z - L/2)^3 \quad \forall z \in [L/2, L]$$

(3.10)

where $S_{01}, S_{11}, \ldots, S_{32}$ are integration constants. It is assumed that both the tension and compression reinforcements of the beam are continuous without any curtailment along its length, leading to the same values of limit elastic and ultimate bending moments. At elastic limit, bending moment reaches its limit value in its absolute

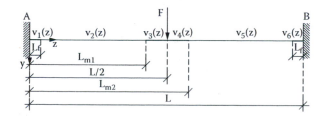

FIGURE 3.3 Fixed beam subjected to central concentrated load.

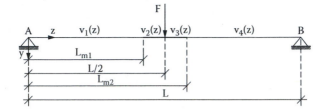

FIGURE 3.4 Simply supported beam subjected to central concentrated load.

terms at sections $z = 0$ and $z = L/2$, simultaneously. By imposing the appropriate equilibrium and compatibility conditions at fixed supports and midspan, integration constants are obtained as

$$S_{01} = S_{11} = 0, S_{21} = \frac{kFL}{16K_E^\phi}, S_{31} = -\frac{kF}{12K_E^\phi}$$

$$S_{02} = \frac{kFL^3}{192K_E^\phi}, S_{12} = 0, S_{22} = -\frac{kFL}{16K_E^\phi}, S_{32} = \frac{kF}{12K_E^\phi}$$

(3.11)

where, k is the load multiplier and F is the point load. Substituting in Equation 3.8, displacement function for the beam in elastic range is determined as

$$\delta_1(z) = \frac{kFz^2}{48K_E^\phi}(3L - 4z) \qquad \forall z \in [0, L/2]$$

[for fixed beam case] (3.12)

$$\delta_2(z) = -\frac{kF(L - 4z)(L - z)^2}{48K_E^\phi} \qquad \forall z \in [L/2, L]$$

Similarly, functions for rotation, bending moment and shear can be readily derived from Equation 3.12. Figure 3.5 shows the bending moment, curvature, rotation and displacement of the beam, plotted along its length; profiles are shown at elastic limit and at collapse as well. It can be seen from the figure that both compatibility and equilibrium conditions are well satisfied. Load multiplier at elastic limit is given by

$$k_e = \frac{8M_E}{FL}$$

(3.13)

At elastic limit, relative rotation, total rotation, and ductility of the hinge formed at midspan are given by

$$\Delta\theta_E^{(midspan)} = \theta_E(z = L_{m1}) - \theta_E(z = L_{m2}),$$

$$\Delta\theta_u^{(midspan)} = \theta_u(z = L_{m1}) - \theta_u(z = L_{m2}),$$

(3.14)

$$\eta = \frac{\Delta\theta_u^{(midspan)}}{\Delta\theta_E^{(midspan)}}$$

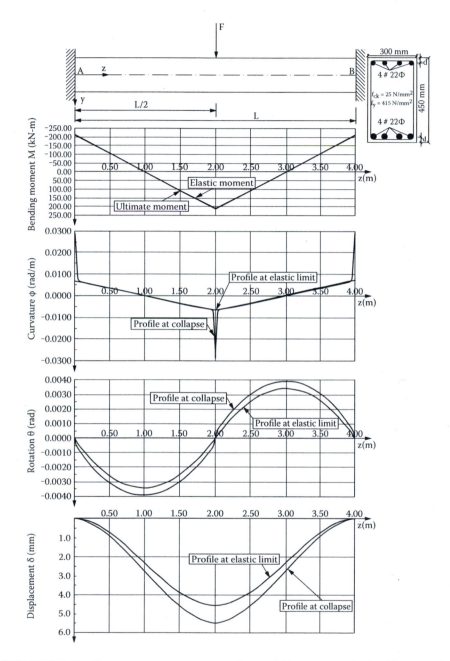

FIGURE 3.5 Bending moment, curvature, rotation, and displacements for fixed beam under central concentrated load.

In terms of moment and curvature-elastic stiffness, relative rotation at elastic limit is given by

$$\Delta\theta_E^{(midspan)} = \frac{M_E\left[L(L_{m1} + 3L_{m2}) - 2\left(L_{m1}^2 + L_{m2}^2\right) - L^2\right]}{K_E^\phi L} \tag{3.15}$$

where, L_{m1} and L_{m2} are the extremities of the plastic hinge, measured along the length of the beam (see Figure 3.3). Similarly, for the plastic hinges formed at fixed supports, relative rotation at elastic limit, total rotation, and ductility ratio are given by

$$\Delta\theta_E^{(support)} = \theta_E(z = L_f), \ \Delta\theta_u^{(support)} = \theta_u(z = L_f), \ \eta = \frac{\Delta\theta_u^{support}}{\Delta\theta_E^{support}} \tag{3.16}$$

where, L_f is the length of plastic hinges formed at the fixed supports. Relative rotation of these hinges can also be expressed as

$$\Delta\theta_E^{(support)} = \frac{M_E L_f(L - 2L_f)}{K_E^\phi L} \tag{3.17}$$

In the elastic-plastic range, the beam is subdivided by six parts (see Figure 3.3). Plastic hinges are present in the first, third, fourth, and sixth parts, while strain remains elastic in the second and fourth parts. Displacement function for each part is given by

$$\delta_1(z) = N_{01} + N_{11}z + N_{21}z^2 + N_{31}z^3 \qquad\qquad \forall z \in [0, L_f]$$

$$\delta_2(z) = N_{02} + N_{12}(z - L_f) + N_{22}(z - L_f)^2 + N_{32}(z - L_f)^3 \qquad \forall z \in [L_f, L_{m1}]$$

$$\delta_3(z) = N_{03} + N_{13}(z - L_{m1}) + N_{23}(z - L_{m1})^2 + N_{33}(z - L_{m1})^3 \qquad \forall z \in [L_{m1}, L/2]$$

$$\delta_4(z) = N_{04} + N_{14}(z - L/2) + N_{24}(z - L/2)^2 + N_{34}(z - L/2)^3 \qquad \forall z \in [L/2, L_{m2}]$$

$$\delta_5(z) = N_{05} + N_{15}(z - L_{m2}) + N_{25}(z - L_{m2})^2 + N_{35}(z - L_{m2})^3 \qquad \forall z \in [L_{m2}, L - L_f]$$

$$\delta_6(z) = N_{06} + N_{16}(z - L + L_f) + N_{26}(z - L + L_f)^2 + N_{36}(z - L + L_f)^3 \ \forall z \in [L - L_f, L]$$

$$\tag{3.18}$$

where $N_{01}, N_{11}, N_{21}, N_{31} \dots, N_{06}, N_{16}, N_{26}, N_{36}$ are integration constants. The functions $\delta_2(z), \delta_5(z)$ are required to satisfy Equation 3.6, while the remaining functions satisfy Equation 3.7. Imposing the respective equilibrium and compatibility conditions, derived are the following set of equations that have to be satisfied:

$$
\begin{cases}
\delta_1(z=0)=0, \quad \varphi_1(z=0)=0, \\
\delta_1(z=L_f)=\delta_2(z=L_f), \phi_1(z=L_f)=\phi_2(z=L_f), M_1(z=L_f) \\
\quad = M_2(z=L_f), V_1(z=L_f)=V_2(z=L_f), \\
\delta_2(z=L_{m1})=v_3(z=L_{m1}), \phi_2(z=L_{m1})=\phi_3(z=L_{m1}), M_2(z=L_{m1})=M_3(z=L_{m1}), \\
V_2(z=L_{m1})=V_3(z=L_{m1}), \delta_3(z=L/2)=v_4(z=L/2), \phi_3(z=L/2)=\phi_4(z=L/2), \\
M_3(z=L/2)=M_4(z=L/2), \ V_3(z=L/2)=V_4(z=L/2)+k\,F, \delta_4(z=L_{m2}) \\
\quad = \delta_5(z=L_{m2}), \\
\phi_4(z=L_{m2})=\phi_5(z=L_{m2}), M_4(z=L_{m2})=M_5(z=L_{m2}), V_4(z=L_{m2})=V_5(z=L_{m2}) \\
\delta_5(z=L-L_f)=\delta_6(z=L-L_f), \ \phi_5(z=L-L_f)=\phi_6(z=L-L_f), \ M_5(z=L-L_f) \\
\quad = M_6(z=L-L_f), \\
V_5(z=L-L_f)=V_6(z=L-L_f), \delta_6(z=L)=0, \quad \phi_6(z=L)=0
\end{cases}
$$

$$(3.19)$$

By solving, integration constants can be determined. By substituting moment as M_E at $(z=L_{m1})$ and M_u at $(z=L/2)$, collapse load multiplier is obtained as

$$
k_c = \frac{8M_u}{FL}; L_f = \frac{L(M_u-M_E)}{4M_u}; L_{m2}-L_{m1} = \frac{L(M_u-M_E)}{2M_u}
$$

$$(3.20)$$

By substituting in Equations 3.15 and 3.17, respectively, relative rotation of plastic hinges formed at midspan and supports, at elastic limit, are obtained as

$$
\Delta\theta_E^{(\text{midspan})} = \frac{\left(M_u^2-M_E^2\right)M_E L}{4K_E^\phi M_u^2} = \frac{\left(M_u^2-M_E^2\right)\phi_E L}{4M_u^2}
$$

$$
\Delta\theta_E^{(\text{support})} = \frac{\left(M_u^2-M_E^2\right)M_E L}{8K_E^\phi M_u^2} = \frac{\left(M_u^2-M_E^2\right)\phi_E L}{8M_u^2}
$$

$$(3.21)$$

In the elastic-plastic range, respective values are given by

$$
\Delta\theta_u^{(\text{midspan})} = \frac{(M_u-M_E)\left[2K_p^\phi M_E + K_E(M_u-M_E)\right]L}{4K_E^\phi K_p^\phi M_u} = \frac{(M_u-M_E)(\phi_u+\phi_E)L}{4M_u}
$$

$$
\Delta\theta_u^{(\text{support})} = \frac{(M_u-M_E)\left[2K_p^\phi M_E + K_E^\phi(M_u-M_E)\right]L}{8K_E^\phi K_p^\phi M_u} = \frac{(M_u-M_E)(\phi_u+\phi_E)L}{8M_u}
$$

$$(3.22)$$

Rotation ductility of the plastic hinges formed at supports and midspan are given by

$$\eta_\theta = \frac{M_u \left[2K_p^\phi M_E + K_E^\phi (M_u - M_E) \right]}{K_p^\phi M_E (M_u + M_E)} = \frac{M_u}{M_u + M_E} \left(\frac{\phi_u + \phi_E}{\phi_E} \right) \tag{3.23}$$

Moment-rotation relationships, in both elastic and plastic zones, are summarized as

$$M(\Delta\theta) = \begin{cases} K_E^\theta \Delta\theta & \Delta\theta \in \left[0, \Delta\theta_E^{(\text{midspan})} \right] \\ M_E + K_P^\theta \left(\Delta\theta - \Delta\theta_E^{(\text{midspan})} \right) & \Delta\theta \in \left[\Delta\theta_E^{(\text{midspan})}, \Delta\theta_u^{(\text{midspan})} \right] \end{cases} \tag{3.24}$$

where the rotational-elastic stiffness and hardening modulus K_E^θ, K_P^θ assume different values for plastic hinges formed at midspan and supports. For hinges formed at midspan and supports, their respective values are given by:

$$K_E^\theta = \left[\frac{4M_u^2}{(M_u^2 - M_E^2)L} \right] K_E^\phi, \quad K_P^\theta = \frac{4K_E^\phi K_P^\phi M_u^2}{(M_u - M_E)(M_E K_P^\phi + M_u K_E^\phi)L} \tag{3.25}$$

$$K_E^\theta = \left[\frac{8M_u^2}{(M_u^2 - M_E^2)L} \right] K_E^\phi, \quad K_P^\theta = \frac{8K_E^\phi K_P^\phi M_u^2}{(M_u - M_E)(M_E K_P^\phi + M_u K_E^\phi)L} \tag{3.26}$$

3.4.2 Simply Supported Beam under Central Concentrated Load

Figure 3.4 shows the simply supported beam under central concentrated load. The beam is examined for increasing design load until collapse; and moment-rotation relationships, in elastic and elastic-plastic ranges, are presented. In the elastic range, the beam is subdivided in two parts whose lengths lie in the range (0, L/2) and (L/2, L), respectively. The displacement functions for both of the parts are given by

$$\delta_1(z) = Y_{01} + Y_{11}z + Y_{21}z^2 + Y_{31}z^3 \qquad \forall z \in [0, L/2]$$
$$\delta_2(z) = Y_{02} + Y_{12}(z - L/2) + Y_{22}(z - L/2)^2 + Y_{32}(z - L/2)^3 \qquad \forall z \in [L/2, L] \tag{3.27}$$

By imposing appropriate equilibrium and compatibility conditions, integration constants of Equation 3.27 are obtained as

$$Y_{01} = 0, \ Y_{11} = \frac{kFL^2}{16K_E^\phi}, Y_{21} = 0, \ Y_{31} = -\frac{kF}{12K_E^\phi}$$

$$Y_{02} = \frac{kFL^3}{48K_E^\phi}, \ Y_{12} = 0, \ Y_{22} = -\frac{kFL}{8K_E^\phi}, \ Y_{32} = \frac{kF}{12K_E^\phi}, \tag{3.28}$$

where k and F are the load multiplier and point load, respectively. Substituting in Equation 3.27, displacement function for the beam in elastic range is determined as:

$$\delta_1(z) = \frac{kFz}{48\,K_E^\phi}(3L^2 - 4z^2) \qquad\qquad \forall z \in [0, L/2]$$

[for simply supported beam]

$$\delta_2(z) = -\frac{kF(L-4z)(L^2 - 8Lz + 4z^2)}{48K_E^\phi} \qquad \forall z \in [L/2, L]$$

$$(3.29)$$

Figure 3.6 shows the bending moment, curvature, rotation, and displacement profiles of the beam, plotted along its length; profiles are shown at elastic limit and collapse as well. It can be seen from the figure that compatibility and equilibrium conditions are well satisfied. The load multiplier at elastic limit is now given by

$$k_e = \frac{4\,M_E}{FL} \qquad\qquad (3.30)$$

Using Equation 3.14, relative rotation of the plastic hinge formed at midspan, at elastic limit, is given by

$$\Delta\theta_E^{(midspan)} = \frac{M_E\left[4LL_{m2} - L^2 - 2\left(L_{m1}^2 + L_{m2}^2\right)\right]}{2K_E^\phi L} \qquad\qquad (3.31)$$

where L_{m1} and L_{m2} are the extremities of the plastic hinge measured along the length of the beam as shown in Figure 3.4 and are given by

$$L_{m1} = \frac{M_E}{M_u}\frac{L}{2}, \quad L_{m2} = L\left(1 - \frac{M_E}{2M_u}\right) \qquad\qquad (3.32)$$

The length of the plastic hinge formed at midspan is given by:

$$L_{m2} - L_{m1} = L\left(1 - \frac{M_E}{M_u}\right) \qquad\qquad (3.33)$$

Substituting in Equation 3.31, relative rotation of the plastic hinge at elastic limit is given by

$$\Delta\theta_E^{(midspan)} = \frac{\left(M_u^2 - M_E^2\right)M_E L}{2K_E^\phi M_u^2} = \frac{\left(M_u^2 - M_E^2\right)\phi_E L}{2M_u^2} \qquad\qquad (3.34)$$

In the elastic-plastic range, the beam is subdivided in four parts (see Figure 3.4). Plastic hinges are present in second and third part, while the strain in the

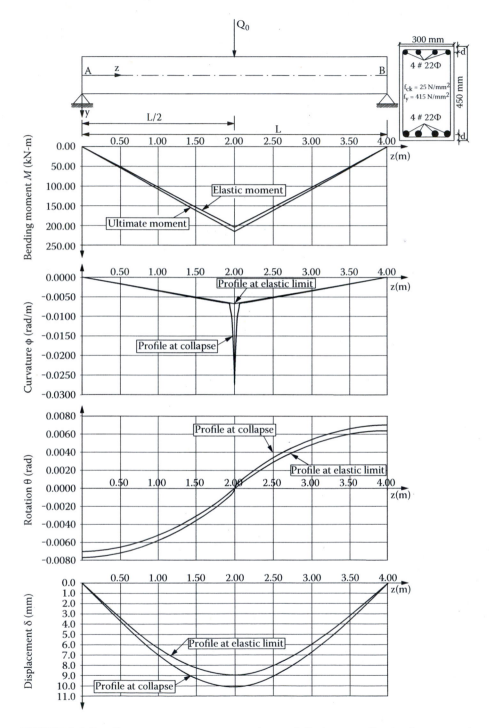

FIGURE 3.6 Bending moment, curvature, rotation, and displacement for simply supported beam under central concentrated load.

first and fourth parts remains elastic. Displacement functions for each part are given by

$$\delta_1(z) = R_{01} + R_{11}z + R_{21}z^2 + R_{31}z^3 \qquad \forall z \in [0, L_{m1}]$$

$$\delta_2(z) = R_{02} + R_{12}(z - L_{m1}) + R_{22}(z - L_{m1})^2 + R_{32}(z - L_{m1})^3 \qquad \forall z \in [L_{m1}, L/2]$$

$$\delta_3(z) = R_{03} + R_{13}(z - L/2) + R_{23}(z - L/2)^2 + R_{33}(z - L/2)^3 \qquad \forall z \in [L/2, L_{m2}]$$

$$\delta_4(z) = R_{04} + R_{14}(z - L_{m2}) + R_{24}(z - L_{m2})^2 + R_{34}(z - L_{m2})^3 \qquad \forall z \in [L_{m2}, L]$$

$$(3.35)$$

where $R_{01}, R_{11}, ..., R_{24}, R_{34}$ are the integration constants that can be determined by imposing the respective equilibrium and compatibility conditions in Equation 3.35. Relative rotation of the plastic hinge formed at midspan is given by

$$\Delta\theta_u^{(midspan)} = \frac{(M_u - M_E)\left[K_E^\phi(M_u - M_E) + 2K_p^\phi M_E\right]L}{2K_E^\phi K_p^\phi M_u} = \frac{(M_u - M_E)(\phi_u + \phi_E)L}{2M_u}$$

$$(3.36)$$

By recalling the relative rotation of the plastic hinge at elastic limit given by Equation 3.34, ductility of the plastic hinge formed at midspan can be expressed as

$$\eta_\theta = \frac{\Delta\theta_u^{(midspan)}}{\Delta\theta_E^{(midspan)}} = \frac{M_u\left[2K_p^\phi M_E + K_E^\phi(M_u - M_E)\right]}{K_p^\phi M_E(M_u + M_E)} = \frac{M_u}{M_u + M_E}\left(\frac{\phi_u + \phi_E}{\phi_E}\right) \qquad (3.37)$$

It is important to note that rotation ductility obtained above is as same as the hinge formed at midspan in the fixed beam, given by Equation 3.25, but the changes in rotational-elastic and hardening modulus are given by

$$K_E^\theta = \left[\frac{2M_u^2}{(M_u^2 - M_E^2)L}\right]K_E^\phi, \quad K_p^\theta = \frac{2K_E^\phi K_p^\phi M_u^2}{(M_u - M_E)(M_E K_p^\phi + M_u K_E^\phi)L} \qquad (3.38)$$

3.4.3 Fixed Beam under Uniformly Distributed Load

In this section, the collapse mechanism and the plastic hinge extension of a fixed RC beam, in bending under increasing uniformly distributed design load until collapse, is examined, and expressions for moment-rotation relationships are derived. A fixed beam of 5 m span considered for the study is shown in Figure 3.7. Two RC beams with cross-sections 300×450 mm and 300×600 mm are analyzed with different percentages of tension and compression reinforcements. The bending moment, curvature, rotation, and deflection function in a closed form in elastic range are given by

$$M_e(z) = \frac{w_0 L^2}{2}\left[-\frac{z^2}{L^2} + \frac{z}{L} - \frac{1}{6}\right] \qquad (3.39)$$

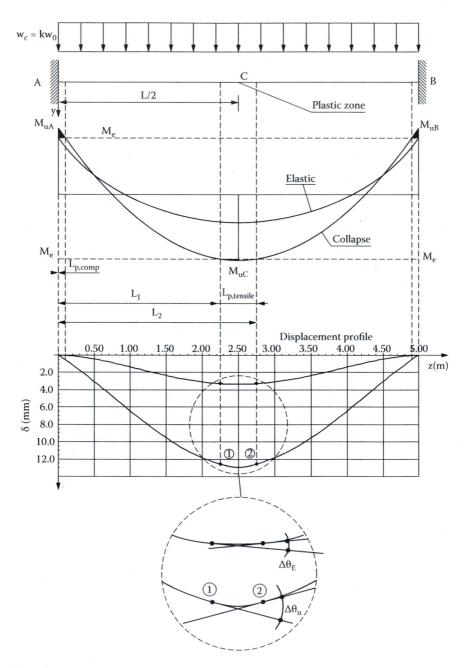

FIGURE 3.7 Fixed beam 300×450 showing position of plastic hinges ($p_t = 2.5\%$; $p_c = 1\%$)

$R_{ck} = 30$; $\sigma_{c0} = 13.228$; $\varepsilon_{c0} = 0.2\%$; $\varepsilon_{cu} = 0.35\%$; $\sigma_y = 380$;

$\sigma_{s0} = 330.435$, $\varepsilon_{s0} = 0.157\%$; $\varepsilon_{su} = 1\%$

$$\phi_e(z) = \frac{M_e(z)}{\bar{K}_E^\phi} = \frac{w_0 L^2}{2\bar{K}_E^\phi}\left[-\frac{z^2}{L^2} + \frac{z}{L} - \frac{1}{6}\right] \tag{3.40}$$

$$\theta_e(z) = \int\frac{M_e(z)}{\bar{K}_E^\phi}dz + G_0 = \frac{w_0 L^3}{12\bar{K}_E^\phi}\left[\frac{3z^2}{L^2} - \frac{2z^3}{L^3} - \frac{z}{L}\right] + G_0 \tag{3.41}$$

$$\delta_e(z) = -\int\theta_e(z)dz + G_1 = \frac{w_0 L^4}{24\bar{K}_E^\phi}\left[\frac{z^2}{L^2} - \frac{2z^3}{L^3} + \frac{z^4}{L^4}\right] + G_1 \tag{3.42}$$

where G_0, G_1 are integration constants depending on boundary kinematical constraints. The equivalent bending stiffness is given by

$$\bar{K}_E^\phi = E_s(p_t + p_c)b(D-d)\frac{D^2}{4} + \frac{2\sigma_{c0}}{\varepsilon_{c0}}\frac{bD^3}{12} \tag{3.43}$$

A procedure based on the static theorem of limit analysis is applied to obtain a lower-bound collapse load multiplier, length of plastic hinge, and relative rotation. It is well known that in the framework of the static theorem, only equilibrium and plastic compatibility conditions are to be fulfilled. For the selected cross-section and reinforcement, let the ultimate moments be M_{uA} and M_{uB} at fixed supports, M_{uC} at midspan, and let M_E be the limit moment (Figure 3.7). Since the load is uniformly distributed, equilibrium requires a parabolic bending moment function over the beam. A statically admissible bending moment distribution $M = M(z)$ is hence a parabola passing through these values at supports and midspan. The equilibrium equations at collapse of these sections are thus given by

$$P_p = \int_q^{x_c} b\sigma_c(\varepsilon_c(y))dy + bq\sigma_{c0} + (A_{sc} - A_{st})\sigma_{s0} = 0 \tag{3.44}$$

$$M_p = \int_q^{x_c} b\sigma_c(\varepsilon_c(y))\left(\frac{D}{2} - y\right)dy + \frac{bq\sigma_{c0}}{2}(D-q) + \sigma_{s0}(A_{st} + A_{sc})\left(\frac{D}{2} - d\right) \tag{3.45}$$

where $q = \frac{\varepsilon_{c,max} - \varepsilon_{c0}}{\varepsilon_{c,max}}x_c$; $\varepsilon_c = \frac{\varepsilon_{c,max}}{x_c}(x_c - y)$. By solving Equation 3.44, depth of neutral axis is obtained as

$$x_c = \frac{3(D-d)(p_t - p_c)\sigma_{s0}\varepsilon_{c,max}}{\sigma_{c0}(3\varepsilon_{c,max} - \varepsilon_{c0})} \tag{3.46}$$

By substituting in Equation 3.45 and imposing the conditions

$$\varepsilon_{sc} = \varepsilon_{s0}$$
$$\varepsilon_{st} = \varepsilon_{s0} \tag{3.47}$$
$$\varepsilon_{c,max} = \varepsilon_{c0}$$

three values of elastic bending moments are obtained, out of which lower bound value is taken as elastic moment, M_E. By further substituting in Equation 3.45 and imposing the condition that $\varepsilon_{c,max} = \varepsilon_{cu}$, the ultimate bending moment of the cross-section is obtained as

$$M_u = b(D-d)\sigma_{s0}\left[Dp_t - d(p_t + p_c) - \frac{3\sigma_{s0}(D-d)(p_c - p_t)^2\left(\varepsilon_{c0}^2 - 4\varepsilon_{c0}\varepsilon_{cu} + 6\varepsilon_{cu}^2\right)}{4\sigma_{c0}(\varepsilon_{c0} - 3\varepsilon_{cu})^2}\right]$$

(3.48)

The relevant uniformly distributed statically admissible load is given by

$$kw_0 = -\frac{d^2M}{dz^2}$$

(3.49)

By integrating the moment distribution along the length of the beam at collapse, we get

$$M(z) = -\frac{kw_0}{2}z^2 + D_1z + D_2$$

(3.50)

Now, by imposing $|M_{uA}| = M_{uC} = M_u$, the collapse load multiplier and integration constants D_1, D_2 are obtained as

$$k = \frac{16M_u}{w_0L^2}, \quad D_1 = \frac{8M_u}{L}, \quad D_2 = -M_u$$

(3.51)

where M_u is given by Equation 3.48. For equilibrium of stresses corresponding to the vanishing value of axial force, the position of neutral axis and linear axial deformation profile can be obtained under the Bernoulli condition for both elastic and elastic-plastic zones. Figure 3.7 shows the details of plastic hinges formed at critical sections. For the zero axial load case in Equation 3.44, strain in concrete in extreme compression fiber is given by

$$\varepsilon_{c,max} = \frac{x_c\varepsilon_{c0}\sigma_{c0}}{3[x_c\sigma_{c0} - (d-D)(p_c - p_t)\sigma_{s0}]}$$

(3.52)

By substituting in Equation 3.44 and equating it to Equation 3.50, depth of neutral axis is obtained as

$$x_c = \frac{b(d-D)(p_c - p_t)\sigma_{s0} + \sqrt{2b}\,F_1(z)}{b\sigma_{c0}}$$

(3.53)

where

$$F_1(z) = \sqrt{-2D_2\sigma_{c0} + z(kw_0z - 2D_1)\sigma_{c0} + 2b(d-D)[-Dp_t + d(p_c + p_t)]\sigma_{c0}\sigma_{s0} - b(d-D)^2\sigma_{s0}^2(p_c - p_t)^2}$$

(3.54)

Subsequently, curvature, rotation, and displacements for the elastic-plastic zone of the beam are given by

$$\phi_p = \frac{\varepsilon_{c,max}}{x_c} = \frac{b\varepsilon_{c0}\sigma_{c0}}{3\sqrt{2b}\ F_1(z)} \tag{3.55}$$

$$\theta_p = \int \phi_p dz + H_0 = \frac{\varepsilon_{c0}\sqrt{b\sigma_{c0}}\ F_1(z)\ln\left[\sigma_{c0}\left(-D_1 + k\ w_0 z + \sigma_{c0}^{-1/2}\sqrt{kw_0}\ F_1(z)\right)\right]}{3\sqrt{2k\ w_0}\ F_1(z)} + H_0 \tag{3.56}$$

$$\delta_p = -\int \theta_p dz + H_1 = \frac{\sqrt{kw_0}\ F_1(z)(-6H_0 kw_0 \sigma_{c0}z + \sqrt{2b}\varepsilon_{c0}\sigma_{c0}F_1(z))}{6k^{3/2}w_0^{3/2}\sigma_{c0}\ F_1(z)} +$$

$$\frac{\sqrt{2\sigma_{c0}}\varepsilon_{c0}F_1(z)[D_1\ln(D_1\sqrt{\sigma_{c0}} - kw_0\sqrt{\sigma_{c0}}z - \sqrt{kw_0}\ F_1(z)) - kw_0 z\ln(-D_1\sigma_{c0} + k\ w_0\sigma_{c0}z + \sqrt{k\ w_0\sigma_{c0}}\ F_1(z))]}{6k^{3/2}w_0^{3/2}\sigma_{c0}\ F_1(z)} + H_1 \tag{3.57}$$

where H_0 and H_1 are integration constants.

By means of the above procedure, functions of curvature, rotation, and displacement of the elastic and elastic-plastic sections of the beam are obtained. Equating rotations and displacements in the connecting points and imposing the conditions zero rotation at midspan and zero displacements at fixed supports, the actual solution for displacement at collapse is determined by solving Equations 3.42 and 3.57. The solution for displacement obtained from the static procedure described above clearly fulfills the continuity requirements at the connection points between the elastic and elastic-plastic parts since the solution presents only a negligible error of 0.005 radians of the rotation at the fixed supports. On the other hand, it is well known that in the framework of static theorem of limit analysis, not all kinematical conditions can be satisfied. The results obtained for the collapse multiplier and plastic hinge length that are intended as relative rotations between the sections whose abscissa are L_1 and L_2, also representing the boundaries of the incoming plastic hinge, are only *lower bounds* of the actual ones. Even though obtained by means of the shown static procedure, the result is nearly close to the exact ones. The proposed modeling for the elastic-plastic beam allows one to obtain the deformation of the beam also with good accuracy. Further, the relative rotations at elastic limit and collapse and ductility ratio (see the enlarged view of Figure 3.7) are given by

$$\Delta\theta_E = \theta_E(z = L_1) - \theta_E(z = L_2)$$

$$\Delta\theta_u = \theta_u(z = L_1) - \theta_u(z = L_2)$$

$$\eta_\theta = \frac{\Delta\theta_u}{\Delta\theta_E} \tag{3.58}$$

The points along the length of the beam, namely, L_1 and L_2 representing extremities of the plastic hinge, can be determined by equating the lower bound value of elastic moment to the actual bending moment given by Equation 3.50 and solving it with respect to variable z. The moment-rotation relationship of the beam in both elastic and plastic zones is summarized as

$$
M = \begin{cases}
\dfrac{K_\phi L^2 \Delta\theta}{(L_1 - L_2)\left[L_1^2 - 3L(L_1 + L_2) + 2(L_1^2 + L_1 L_2 + L_2^2)\right]} & \Delta\theta \in [0, \Delta\theta_E] \\[4mm]
M_E + \left(\dfrac{M_u - M_E}{\Delta\theta_u - \Delta\theta_E}\right)(\Delta\theta - \Delta\theta_E) & \Delta\theta \in [\Delta\theta_E, \Delta\theta_u]
\end{cases}
\tag{3.59}
$$

3.5 NUMERICAL STUDIES AND DISCUSSIONS

The above developed procedure is verified with numerical examples. RC beams of cross-section 300×450 mm, reinforced with 4# 22 Φ in both tensile and compression zone, under central point load are now examined. Two support conditions are considered: (1) both ends fixed and (2) both ends simply supported. The spans of the beams are varied as 3.5 m, 4 m, 4.5 m, and 5 m. The spans are selected in a close range, since the objective is to examine their influence on moment-rotation relationship; and at the same time, the chosen cross-section shall be also accommodated. Further, the cross-section and the reinforcement of the beams (with different spans) are kept the same so that the influence of their variation on moment-curvature is controlled as the moment-rotation relationship is derived from the bilinear approximation of moment-curvature.

The percentages of reinforcements, both in tension and compression ($p_t = p_c = 1.21\%$), are kept less than the balanced section ($p_{t,bal} = 1.85\%$) to initiate a tensile failure in the beam. Axial force is varied as (1) 0 kN, (2) 100 kN, and (3) 200 kN but kept well below the critical load value (p^* for the chosen cross-section is 291.51 kN), since the load level in closer proximity to the critical may indicate the influence of compression failure. Moment-rotation curves are plotted only for the plastic hinge formed at the midspan. However, details of plastic hinges formed at the support can be seen from Tables 3.1 and 3.2 for fixed beam and simply supported beam, respectively.

Figures 3.8 and 3.9 show the moment rotation of the fixed beam and simply supported beam under different axial forces, respectively. The curves in these figures are plotted for the beams of 4 m span only, but Tables 3.1 and 3.2 show the details for the beams with different spans considered in the study. The tables and Figures 3.8 and 3.9 show that for the fixed beam and simply supported beams of specific span length, say L m, the rotational-elastic stiffness K_E^θ decreases for the increase in axial force while rotational-hardening modulus K_p^θ increases; this is true for beams of all spans examined in the study. For the same axial force, increase in span length of the fixed and simply supported beams leads to the decrease of both rotational-elastic and hardening modulus. Relative rotation of plastic hinges formed at the supports (in the case of fixed beam) and at the midspan increases with the increase in span length

TABLE 3.1

Hinge Properties for a Fixed Beam

Description of Data and Properties

		(Fixed Beam)				Hinge at Support						Hinge at Midspan					
L (m)	P (kN)	ϕ_E (rad/m)	ϕ_u (rad/m)	M_E (kN-m)	M_u (kN-m)	$\Delta\theta_E$ (rad)	$\Delta\theta_u$ (rad)	η	L_p (mm)	K_E^θ (kNm/rad)	K_p^θ (kNm/rad)	$\Delta\theta_E$ (rad)	$\Delta\theta_u$ (rad)	η	L_p (mm)	K_E^θ (kN-m/rad)	K_p^θ (kN-m/rad)
3.50	0	0.0068	0.0288	206.65	214.34	0.000210	0.000559	2.665	31.41	985272	22025	0.000419	0.001118	2.665	62.82	492636	11013
3.50	100	0.0072	0.0294	222.97	233.29	0.000272	0.000708	2.605	38.70	820555	23660	0.000543	0.001416	2.605	77.40	410278	11830
3.50	200	0.0076	0.0300	238.89	252.07	0.000338	0.000860	2.545	45.77	707097	25252	0.000676	0.001720	2.545	91.53	353549	12626
4.00	0	0.0068	0.0288	206.65	214.34	0.000240	0.000639	2.665	35.89	862113	19272	0.000479	0.001278	2.665	71.79	431057	9636
4.00	100	0.0072	0.0294	222.97	233.29	0.000311	0.000809	2.605	44.23	717986	20702	0.000621	0.001618	2.605	88.46	358993	10351
4.00	200	0.0076	0.0300	238.89	252.07	0.000386	0.000983	2.545	52.31	618710	22096	0.000772	0.001966	2.545	104.61	309355	11048
4.50	0	0.0068	0.0288	206.65	214.34	0.000270	0.000719	2.665	40.38	766323	17131	0.000539	0.001438	2.665	80.76	383161	8565
4.50	100	0.0072	0.0294	222.97	233.29	0.000349	0.000910	2.605	49.76	638210	18402	0.000699	0.001820	2.605	99.52	319105	9201
4.50	200	0.0076	0.0300	238.89	252.07	0.000434	0.001106	2.545	58.84	549964	19640	0.000869	0.002211	2.545	117.69	274982	9820
5.00	0	0.0068	0.0288	206.65	214.34	0.000300	0.000799	2.665	44.87	689690	15418	0.000599	0.001597	2.665	89.74	344845	7709
5.00	100	0.0072	0.0294	222.97	233.29	0.000388	0.001011	2.605	55.29	574389	16562	0.000776	0.002022	2.605	110.58	287194	8281
5.00	200	0.0076	0.0300	238.89	252.07	0.000483	0.001229	2.545	65.38	494968	17676	0.000965	0.002457	2.545	130.76	247484	8838

TABLE 3.2
Details of Hinges Formed at Midspan in a Simply Supported Beam with a Central Point Load

		Description of Data and Properties (Simply Supported Beam)							Hinge at Midspan		
L (m)	P (kN)	ϕ_ε (rad/m)	ϕ_u (rad/m)	M_ε (kN-m)	M_u (kN-m)	$\Delta\theta_\varepsilon$ (rad)	$\Delta\theta_u$ (rad)	η	L_p (mm)	K_ε^θ (kN-m/rad)	K_p^θ (kN-m/rad)
3.50	0.00	0.0068	0.0288	206.65	214.34	0.000839	0.002236	2.665	125.63	246318	5506
3.50	100.00	0.0072	0.0294	222.97	233.29	0.001087	0.002831	2.605	154.81	205139	5915
3.50	200.00	0.0076	0.0300	238.89	252.07	0.001351	0.003440	2.545	183.07	176774	6313
4.00	0.00	0.0068	0.0288	206.65	214.34	0.000959	0.002556	2.665	143.58	215528	4818
4.00	100.00	0.0072	0.0294	222.97	233.29	0.001242	0.003236	2.605	176.92	179496	5176
4.00	200.00	0.0076	0.0300	238.89	252.07	0.001544	0.003931	2.545	209.22	154677	5524
4.50	0.00	0.0068	0.0288	206.65	214.34	0.001079	0.002875	2.665	161.52	191581	4283
4.50	100.00	0.0072	0.0294	222.97	233.29	0.001397	0.003640	2.605	199.04	159552	4601
4.50	200.00	0.0076	0.0300	238.89	252.07	0.001737	0.004423	2.545	235.37	137491	4910
5.00	0.00	0.0068	0.0288	206.65	214.34	0.001199	0.003195	2.665	179.47	172423	3854
5.00	100.00	0.0072	0.0294	222.97	233.29	0.001553	0.004045	2.605	221.15	143597	4140
5.00	200.00	0.0076	0.0300	238.89	252.07	0.001931	0.004914	2.545	261.53	123742	4419

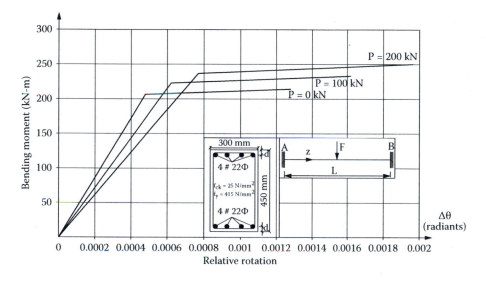

FIGURE 3.8 Moment-rotation of fixed beam under different axial forces:

$d = 30$; $R_{ck} = 25$; $\sigma_{c0} = 13.228$; $\varepsilon_{c0} = 0.2\%$; $\varepsilon_{cu} = 0.35\%$;

$\sigma_y = 415$; $\sigma_{s0} = 330.435$, $\varepsilon_{s0} = 0.157\%$; $\varepsilon_{su} = 1\%$

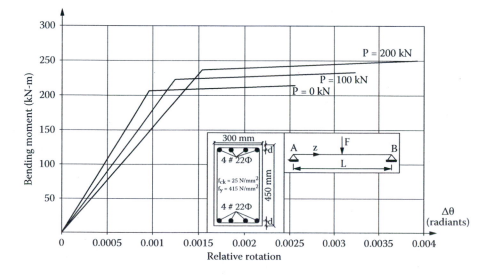

FIGURE 3.9 Moment-rotation of simply supported beam under different axial forces:

$d = 30$; $R_{ck} = 25$; $\sigma_{c0} = 13.228$; $\varepsilon_{c0} = 0.2\%$; $\varepsilon_{cu} = 0.35\%$;

$\sigma_y = 415$; $\sigma_{s0} = 330.435$, $\varepsilon_{s0} = 0.157\%$; $\varepsilon_{su} = 1\%$

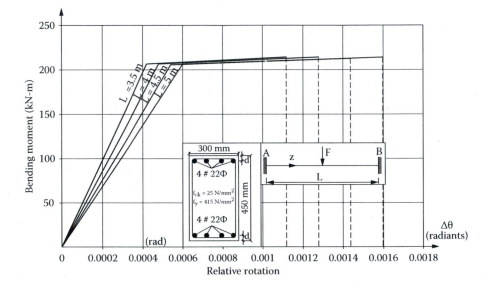

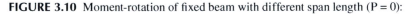

FIGURE 3.10 Moment-rotation of fixed beam with different span length (P = 0):

$d = 30$; $R_{ck} = 25$; $\sigma_{c0} = 13.228$; $\varepsilon_{c0} = 0.2\%$; $\varepsilon_{cu} = 0.35\%$; $\sigma_y = 415$;

$\sigma_{s0} = 330.435$, $\varepsilon_{s0} = 0.157\%$; $\varepsilon_{su} = 1\%$

for the same axial force level; this is true for both at elastic and ultimate stages. Also, for the same span length, say for L m, increase in the axial force increases the relative rotation of the hinges, both at elastic and ultimate stages. Ductility ratios of plastic hinges formed at supports (in case of fixed beams only) and midspan are the same for beams of all spans, subjected to the same axial force; they decrease with the increase in the axial force. The length of plastic hinges formed, both at supports and midspan, increases with the increase in axial force for the same span of the beams. Figures 3.10 and 3.11 show moment-rotation curves of plastic hinges formed at midspan of the fixed beam and simply supported beams of cross-section 300 × 450 mm, with different spans under consideration; plots show the behavior for no axial force. It can be seen from the figures and tables that ductility ratios of plastic hinges formed at supports (in the case of fixed beams) and midspan are the same even with the increase in the span length, under the same axial force. However, there is increase in the length of these plastic hinges with the increase in span lengths, both those formed at supports and those at midspan as well; the length of plastic hinges formed at midspan is double those formed at the supports in the case of fixed beams examined.

Figures 3.12 and 3.13 show the moment-rotation curves for the beam of 4 m span, with two different percentages of tensile reinforcements: (1) 1.21%, which is the same as p_c, and (2) 1.85%, which is the same as $p_{t,bal}$. It is seen from the figures that increase in tension reinforcement certainly increases relative rotations at elastic and

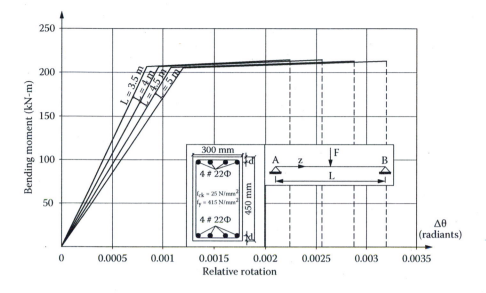

FIGURE 3.11 Moment-rotation of simply supported beam with different span length (P = 0):

$$d = 30; \ R_{ck} = 25; \ \sigma_{c0} = 13.228; \ \varepsilon_{c0} = 0.2\%; \ \varepsilon_{cu} = 0.35\%; \ \sigma_y = 415;$$

$$\sigma_{s0} = 330.435; \ \varepsilon_{s0} = 0.157\%; \ \varepsilon_{su} = 1\%$$

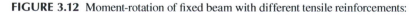

FIGURE 3.12 Moment-rotation of fixed beam with different tensile reinforcements:

$$d = 30; \ R_{ck} = 25; \ \sigma_{c0} = 13.228; \ \varepsilon_{c0} = 0.2\%; \ \varepsilon_{cu} = 0.35\%; \ \sigma_y = 415;$$

$$\sigma_{s0} = 330.435; \ \varepsilon_{s0} = 0.157\%; \ \varepsilon_{su} = 1\%$$

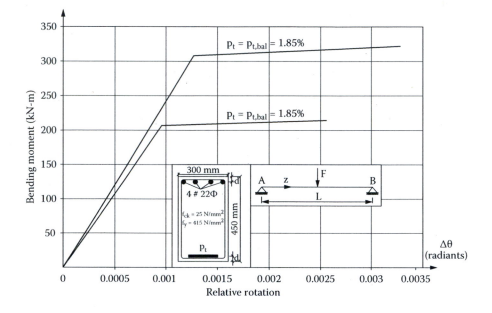

FIGURE 3.13 Moment-rotation of simply supported beam with different tensile reinforcements:

$$d = 30; \, R_{ck} = 25; \, \sigma_{c0} = 13.228; \, \varepsilon_{c0} = 0.2\%; \, \varepsilon_{cu} = 0.35\%; \, \sigma_y = 415;$$

$$\sigma_{s0} = 330.435; \, \varepsilon_{s0} = 0.157\%; \, \varepsilon_{su} = 1\%$$

collapse states, and it also increases moments at elastic and ultimate stages. It is interesting to note that ductility increases only to a marginal extent in comparison with the increases in moments and relative rotations. Therefore, members of RC frames located in seismic zones shall be designed with lesser percentage of tensile reinforcement (which, in other words, is not a compromise on ductility) in comparison with the balanced section, as this can certainly ensure a tensile failure; savings in steel can be seen since a derived benefit apart from ensuring the required ductility. Also, fixing the percentage of compression reinforcement, either equal to that of tension steel or lesser, will be advantageous.

Figures 3.14 and 3.15 show the moment-rotation plots for fixed beams (under uniformly distributed load) of two cross-sections, 300 × 450 mm and 300 × 600 mm, respectively. Table 3.3 shows the moment-rotation and ductility ratio for example cases considered. It is seen that the ductility ratio considerably increases for beams with tension failure compared with that of compression failure, showing also a reduction in length of plastic hinge thus formed. It is also observed that there is a reduction in the length of plastic hinge and increase in ductility ratio when the percentage of tension reinforcement decreases. The required ductility shall be fixed on the basis of demand capacity ratio of the building frame under earthquake loads obtained from preliminary assessment and appropriate input parameter, namely, (1) the section causing tension failure or (2) a balanced section can be chosen.

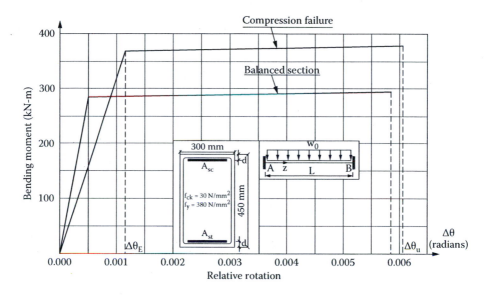

FIGURE 3.14 Moment-rotation of RC beam 300×450:

$d = 30$; $R_{ck} = 30$; $\sigma_{c0} = 13.228$; $\varepsilon_{c0} = 0.2\%$; $\varepsilon_{cu} = 0.35\%$; $\sigma_y = 380$;

$\sigma_{s0} = 330.435$; $\varepsilon_{s0} = 0.157\%$; $\varepsilon_{su} = 1\%$

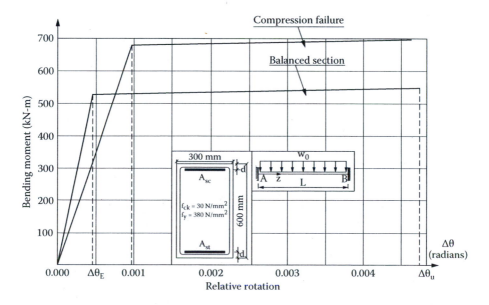

FIGURE 3.15 Moment-rotation of RC beam 300×600:

$d = 30$; $R_{ck} = 30$; $\sigma_{c0} = 13.228$; $\varepsilon_{c0} = 0.2\%$; $\varepsilon_{cu} = 0.35\%$; $\sigma_y = 380$;

$\sigma_{s0} = 330.435$; $\varepsilon_{s0} = 0.157\%$; $\varepsilon_{su} = 1\%$

TABLE 3.3

Details of Hinges Formed at Midspan and Supports in a Fixed Beam under Uniformly Distributed Load

Description Classification	Section 300×450		Section 300×600	
	Compression Failure	Balanced	Compression Failure	Balanced
$P_t (\%)$	2.50	1.84	2.50	1.84
$P_c (\%)$	1.00	1.00	1.00	1.00
$\Delta\theta_E$ (rad)	0.00112	0.00051	0.00096	0.00046
$\Delta\theta_u$ (rad)	0.00606	0.00586	0.0046	0.00475
η	5.19	11.60	4.81	10.35
M_E (Nm)	3.67E+05	2.842E+05	6.797E+05	5.295E+05
M_u (Nm)	3.75E+05	2.938E+05	6.975E+05	5.472E+05
k	4.789	3.756	8.899	6.99
$(\ell_p)_{support}$ (mm)	12.30	2.95	14.03	4.0
$(\ell_p)_{midspan}$ (mm)	495.0	242.91	529.1	283.7

3.6 CONCLUSIONS

A detailed methodology for estimating moment-rotation for RC sections in both elastic and plastic zones separately is presented. Nonlinear characteristics of constitutive materials, namely, concrete and reinforcing steel, according to Eurocode currently in prevalence are considered, while deriving the analytical expressions for moment-rotation in a closed form. The expressions are derived from the earlier developed moment-curvature relationship in Chapter 2, considering a bilinear modeling. Collapse mechanism and plastic hinge extension of RC beams in bending under increasing design load until collapse is presented. Axial forces limiting to result in tensile failure of the chosen beams are also considered during the analysis. The proposed moment-rotation relationships are verified for equilibrium and compatibility conditions and ductility ratios of fixed beams and simply supported beams of different span length are presented. The bilinear approximation of moment-curvature used in the study does not seem to influence the derived analytical expressions of moment-rotation relationships, thus providing a reasonably accurate estimate.

Rotational-elastic stiffness and hardening modulus, the main contributors of ductility, are influenced by the span length of the beam, axial force level, and percentage of steel reinforcement of the cross-section but are not influenced by the support constraints. Ductility is not influenced by the span length of fixed and simply supported beams and their support constraints, whereas length of the plastic hinges is influenced significantly. It is advantageous to limit the value of tension reinforcement less than that required for a balanced section because ductility is not being influenced by the increase in percentage of tension reinforcement (for a fixed value of compression reinforcement). This initiates tensile failure for certain and makes use of the member ductility for effective redistribution of moments at sections where hinges are formed. In the present context of increased emphasis on ductile detailing and displacement-based design approach for structures under seismic loads, analytical estimates of moment-rotation, ductility, and length of plastic hinges presented in a closed form can be seen as useful contributions. It is reemphasized that ductility estimates of RC sections should be made with caution in the presence of axial force. Estimates of critical axial force provided in the closed form can also be useful in this context.

The study verified some of the important facts through proposed analytical expressions presented in a closed form. They are useful for designing special moment-resisting RC framed structures, where ductility is an important design parameter. The method verifies a safe seismic design procedure and can be useful for the practicing engineers as well.

3.7 SPREADSHEET PROGRAM

The spreadsheet program used to estimate the moment-rotation relationship simplifies the complexities involved in such an estimate, thus encouraging the practicing structural designers to use it instantly and with confidence. A compact disc with relevant contents can be downloaded free from the following Web site: http://www.crcpress.com/e_products/downloads/download.asp?cat_no=K10453.

3.7.1 STEP-BY-STEP PROCEDURE TO USE THE NUMERICAL METHOD ON THE WEB SITE

Using the same procedure as explained in Section 2.10, moment-curvature for the chosen cross-section is first determined. The program based on the numerical procedure automatically computes the moment-rotation for the RC beam, using bilinear approximation of the moment-curvature, thus obtained. A sample case problem is presented for fixed beam and a simply supported beam with 3.5 m span length, with different point loads. Please note that the hypothesis discussed above is verified for a tensile failure only. To ensure the failure as tensile, you will cross-check two important parameters: (1) axial load not to exceed P^* value (given by Equation 3.2), as well as (2) the percentage of tension reinforcement not to exceed $P_{t,bal}$.

4 Bounds for Collapse Loads of Building Frames Subjected to Seismic Loads

A Comparison with Nonlinear Static Pushover

4.1 SUMMARY

Recent updates of international codes on seismic analysis and design of buildings reflect the threats to existing buildings under more frequent earthquakes foreseen in the near future. The objective of ensuring structural safety of these buildings under seismic action intensifies their performance assessment for which pushover analysis is widely accepted as a rapid and reasonably accurate method. However, approaches based on limit analysis procedures (both static and kinematic theorems of plasticity theory) have also been equally popular for addressing issues related to structural safety in situations of extreme loads that can jeopardize buildings and could threaten the lives of inhabitants. A comparison between the forecast of design base shear obtained by pushover analysis and collapse loads based on limit analysis procedures is advantageous to establish confidence in the obtained results. In this chapter, we discuss the analytical procedures to determine the collapse loads by limit analysis and pushover as well. Comparison of the results obtained by employing the above tools on multistory moment-resisting reinforced concrete frames subjected to seismic loads is presented. Displacement-controlled pushover analysis is performed on the building frames whose input parameters like axial force–bending moment yield interaction and moment-rotation are derived based on the detailed mathematical modeling presented in earlier chapters. Bounds for collapse loads based on both static and kinematic theorems of limit analysis are obtained using mathematical programming tools. Computer code used to determine the collapse multipliers is given in Chapter 6.

Numerical studies conducted show that the design base shear computed using nonlinear static pushover, for an accepted level of damage like collapse prevention, predicts the response value closer to the upper bounds obtained by plasticity theorems, in certain cases considered. The proposed bounds for collapse loads obtained in closed form, which fit with pushover analysis to a good accuracy, become a useful tool for preliminary design and assessment as well. This study helps the designers and researchers to use displacement-controlled pushover analysis with improved confidence as their results of different examples are compared with other similar methods used to assess the collapse loads. While pushover analysis is recommended as an appropriate tool for seismic assessment of buildings, it is emphasized that accuracy of pushover depends on characteristic inputs presented in the earlier chapters, and design base shear will be better estimated using the proposed expressions.

4.2   INTRODUCTION

The increased use of concrete as the primary structural material in several complex structures such as reactor vessels, dams, offshore structures, and the like needs an accurate estimate of this material response when subjected to a variety of loads that determine the presence of bending, shear, and axial force (Abu-Lebdeh and Voyiadjis 1993; Paulay and Priestley 1992). Seismic design philosophy demands energy dissipation/absorption by postelastic deformation for collapse prevention during major earthquakes. Most of the existing RC buildings do not comply with revised seismic codes as a result of material degradation with age, as well as increase in seismic intensity imposing higher design loads. In such situations, performance assessment of existing buildings becomes inevitable to estimate their structural safety. While Gilbert and Smith (2006) showed a parameter-varying approach to identify constitutive nonlinearities in structures subjected to seismic excitations, the objective of ensuring safe buildings intensifies the above-stated concerns for which nonlinear static pushover analysis (NLSP) can be seen as a rapid and reasonably accurate method (Esra and Gulay 2005). Pushover analysis accounts for inelastic behavior of building models and provides reasonable estimates of deformation capacity while identifying critical sections likely to reach limit state during earthquakes (Chopra and Goel 2000). A qualitative insight of input parameters required for performing nonlinear static pushover is presented in earlier chapters. In this chapter, collapse multipliers of RC building frames with different geometry are assessed by employing different procedures, namely, (1) displacement-controlled nonlinear static pushover; (2) upper bound, or kinematic theorem; (3) lower bound, or static theorem; and (4) step-by-step load increment procedure by employing the force-controlled method. The results obtained are then compared.

4.3   COLLAPSE MULTIPLIERS

In this section, the procedure employed for obtaining the collapse load multipliers on RC building frames is briefly presented. For the sake of simplicity, a regular frame with m spans and n floors is considered. Let L be the length of all floor beams and H be the height of all floors. Let Q_0 be the constant vertical load on beams at midspan, corresponding to the sum of dead loads and appropriate live loads (IS 1893, 2002; Chopra 2003; Chandrasekaran and Roy 2006). Let F_n, F_{n-1}, ... F_i, ... F_2, F_1 be the set of transverse forces distributed along the height of the building for the base shear

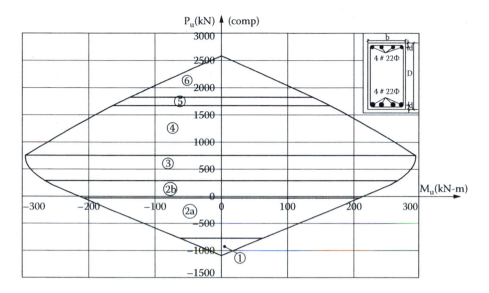

FIGURE 4.1 P-M interaction curve for RC beam

$d = 30$; $b = 300$; $D = 450$; $R_{ck} = 25$; $\sigma_{c0} = 11.023$; $\varepsilon_{c0} = 0.2\%$; $\varepsilon_{cu} = 0.35\%$;

$\sigma_y = 415$; $\sigma_{s0} = 360.87$, $\varepsilon_{s0} = 0.172\%$; $\varepsilon_{su} = 1\%$

computed from the code (IS 1893, 2002). They are assumed to act at every floor level where a constant, equal mass is lumped.

$$F_i = V_b \frac{W_i\,(i\,H)^2}{\displaystyle\sum_{i=1}^{n} W_i (i\,H)^2} \quad \forall i \in \{1,0,...,n\} \tag{4.1}$$

where V_b is the base shear and W_i is the seismic weight of the floors computed from dead load and percentage of appropriate live loads as specified in the code (IS 1893, 2002). All beams and columns are considered to have the same ultimate bending strength, $M_{u,b}$ and $M_{u,c}$, respectively. Without loss of generality, only the cases of weak or balanced section for beams are considered, while columns are considered to be strongly reinforced. In the following section, a straightforward procedure for obtaining upper bounds, K_k (using kinematic theorem), and lower bounds, K_s (using static theorem), of the collapse multiplier is proposed. Figures 4.1 and 4.2 show the P-M interaction of the RC beam and the column, respectively. Figures 4.3 and 4.4 show the moment-rotation capacity of the tensile and compressive plastic hinges, which are used for the analysis. For any other RC section, the reader can easily determine these input parameters either using the enclosed CD or referring to the explicit expressions given in Chapters 1 and 3 of *Seismic Design Aids*. It is well known that the limit analysis theorem is applicable to convex domains only where the normality

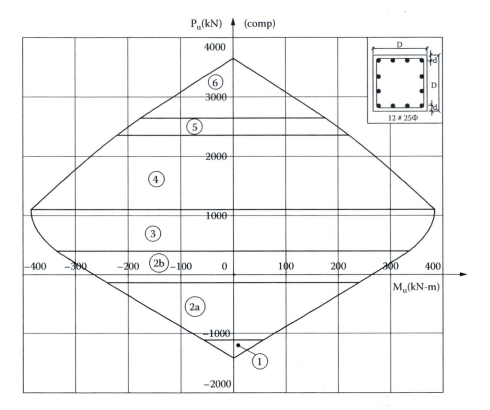

FIGURE 4.2 P-M interaction curve for RC column (450×450)

$$d = 40; \quad R_{ck} = 25; \quad \sigma_{c0} = 11.023; \quad \varepsilon_{c0} = 0.2\%; \quad \varepsilon_{cu} = 0.35\%;$$

$$\sigma_y = 415; \quad \sigma_{s0} = 360.87, \quad \varepsilon_{s0} = 0.172\%; \quad \varepsilon_{su} = 1\%$$

rule is verified. A detailed insight of verification of flow rule for the proposed P-M interaction domain is presented in Chapter 5.

4.3.1 KINEMATIC MULTIPLIER, K_k

The proposed upper-bound collapse multiplier, K_k, for the seismic design forces distribution shown in Equation 4.1 is obtained by means of the kinematical procedure of limit analysis. This is based on the assumption of a failure mode shown in Figure 4.5, fulfilling only the compatibility requirement that allows the evaluation of the total dissipation.

$$\sum_{j=1}^{p} M_{u,j}\, \Delta\theta_j - \sum_{i=1}^{n} K_k\, F_i \cdot \delta_{h,i} - \sum_{i=1}^{n}\sum_{k=1}^{m} Q_0\, \delta_{v,ik} \geq 0 \qquad (4.2)$$

where M_u is the ultimate bending moment of the element considered, p is the number of plastic hinge, n is the number of floors, m is number of spans, L is the length of

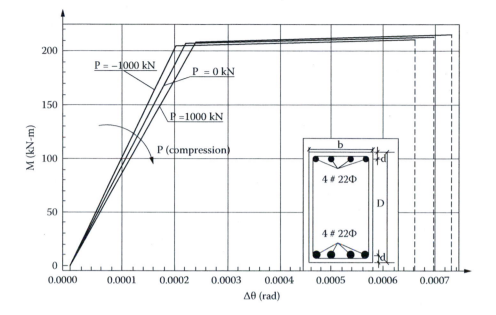

FIGURE 4.3 Moment-rotation for RC beams (tensile hinge) ($p_t = 1.207\%$; $p_c = 1.207\%$)

$b = 300$; $d = 30$; $D = 450$; $R_{ck} = 25$; $\sigma_{c0} = 11.023$; $\varepsilon_{c0} = 0.2\%$; $\varepsilon_{cu} = 0.35\%$;

$\sigma_y = 415$; $\sigma_{s0} = 360.87$; $\varepsilon_{s0} = 0.172\%$; $\varepsilon_{su} = 1\%$

beams, Q_0 is the concentrated load on the beam at midspan, $\Delta\theta_i$ is the relative rotation rate, δ_{hi} is the floor displacement rate in the horizontal direction and δ_{vk} is the beam displacement rate in the vertical direction. The searched kinematical multiplier is given by

$$K_k = \frac{\sum\limits_{j=1}^{p} M_{u,j}\,\Delta\theta_j - \sum\limits_{i=1}^{n}\sum\limits_{k=1}^{m} Q_0\,\delta_{v,ik}}{\sum\limits_{i=1}^{n} F_i \cdot \delta_{h,i}} \geq K_c \qquad (4.3)$$

The simplest failure mode is assumed corresponding to the positioning of plastic hinges at critical sections, namely, all beam supports and the bottom section of first-floor columns. The modeled failure mode assumes point-wise plastic hinges at which relative plastic rotations occur. For this failure mode, vertical loads do not work, and hence the revised kinematical multiplier is given by

$$K_k = \frac{\sum\limits_{j=1}^{p} M_{u,j}}{\sum\limits_{i=1}^{n} F_i \cdot i \cdot H} = \frac{2(1+2n)[2\,m\,n\,M_{u,b} + (m+1)\,M_{u,c}]}{3H\,n\,(n+1)\,V_b} \geq K_c \qquad (4.4)$$

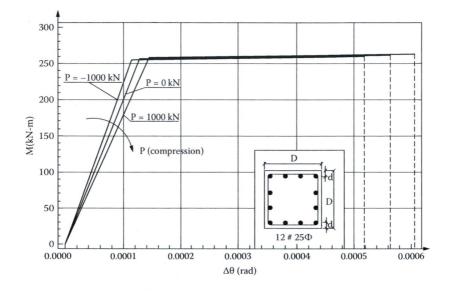

FIGURE 4.4 Moment-rotation for RC column (compression hinge) ($p_t = 1.06\%$; $p_c = 1.06\%$)

$b = 450$; $d = 40$; $D = 450$; $R_{ck} = 25$; $\sigma_{c0} = 11.023$; $\varepsilon_{c0} = 0.2\%$; $\varepsilon_{cu} = 0.35\%$;

$\sigma_y = 415$; $\sigma_{s0} = 360.87$; $\varepsilon_{s0} = 0.172\%$; $\varepsilon_{su} = 1\%$

4.3.2 STATIC MULTIPLIER, K_s

A static multiplier, K_s, constituting a lower bound of the collapse multiplier, is to be obtained by employing the static procedure of limit analysis, based on the search of a statically admissible stress distribution. While the stress field fulfilling only equilibrium equations must be contained in the ultimate strength limits, no kinematic compatibility equations, in elastic or plastic range, are required to be satisfied. Statically admissible distribution of bending moment at any section is considered, and its distribution is set to satisfy the condition that bending moment is less than or equal to ultimate bending moment at the cross-section. Equilibrium equations written for various characteristic sections of the structure and satisfactory conditions for plastic compatibility at these sections impose constraints to the mathematical programming problem (Rustem 2006; Yakut, Yilmaz, and Bayili 2001). The static theorem of limit analysis enables one to compute the collapse static multiplier of loads, K_s, satisfying the following relationship:

$$K_c = \max(K_s), \tag{4.5}$$

where K_c is the collapse multiplier to be bounded. The usual hypotheses of piecewise-linear structure having characteristics of piecewise-constant geometry and strength, subjected to concentrated loads and convex yield domain with plane boundaries, are applied (Nunziante and Ocone 1988). Thus, the associated plastic flow rule for small displacements simplifies the procedure, in static instance, to a problem of optimal research by means of linear programming. In order to give an idea of the computational

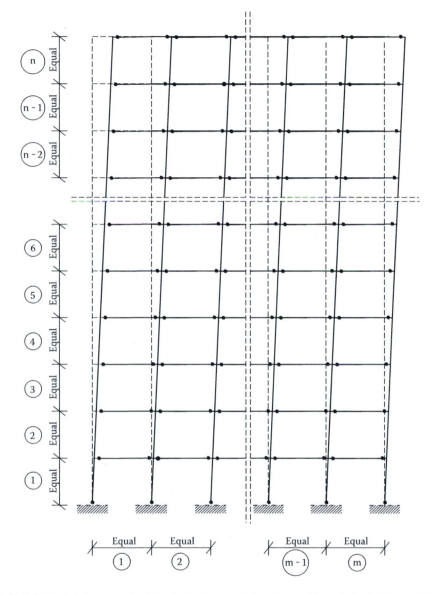

FIGURE 4.5 RC frame under lateral displacement showing position of plastic hinges (kinematic theorem).

tasks required to fulfill the above general procedure, we shall evaluate the number of equations and variables involved in the study of an ordinary rectangular mesh frame. For [n] floors and [m] spans subjected to central concentrated load [Q_0], the number of characteristic sections is [n(5m + 2)]. The number of redundancies become [3mn] and the number of independent equilibrium equations become [2n (m + 1)], making the number of variables in the problem, represented by the redundant moments, [3mn]. By using monodimensional strength domains for beams and columns (plasticization caused only due to bending moment and P-M interaction is ignored), the number of plastic compatibility inequalities becomes [n(10m + 4)]. Plastic compatibility inequalities at midspan and at supports of the beams are given by

$$- |M_{u,b}| \le M_{i,k} \le |M_{u,b}| \qquad \forall k \in \{0,1,2,...,m\}, \ \forall i \in \{0,1,2,...,n\} \qquad (4.6)$$

Further, inequalities at the column supports are given by

$$- |M_{u,c}| \le M_{i,k} \le |M_{u,c}| \qquad \forall k \in \{0,1,2,...,m+1\}, \ \forall i \in \{0,1,2,...,n\} \qquad (4.7)$$

Thus, the total number of equations and inequalities amounts to [6n(2m + 1)]. By solving the linear programming problem using LINGO (Raphel, Marak, and Truszcynski 2002; Sforza 2002) characterized by these equations and inequalities, static multiplier can be determined. One can foresee the complexities involved in establishing the above equilibrium equations and inequalities, for a multistory building frame, in particular.

An approximate and simplified procedure is therefore desirable to determine the collapse multiplier by overcoming the above-mentioned complexities. A statically admissible solution is obtained as the sum of results of two cases, namely, (1) the solution corresponding to vertical concentrated loads on beams causing linear bending moment diagram, satisfying null moments at supports; and (2) the solution corresponding to the distribution of floor shear equally to (m + 1) floor columns assuming null moments at the column center and obtaining end moments at the *i*th floor. In the latter case, frame node equilibrium is fulfilled by equating the end moments of columns with that of beams. Figure 4.6 shows the bending moment diagrams for the two cases mentioned above. At the extreme joints of beams, bending moment is equal to the sum of end moments of columns from upper and lower floors, while at internal nodes, two adjacent beams share this value. The sum of the equilibrated bending moment distributions, $[K_s (M_F + M_Q)]$ (the subscript $_F$ stands for floor shear, and $_Q$ stands for vertical concentrated load) shall satisfy the static compatibility conditions given by Equations 4.6 and 4.7. However, kinematic compatibility at nodes of the frame is not satisfied, and hence the obtained multiplier is only a static lower bound of the collapse multiplier K_c. It is interesting to verify that for a strong column–weak beam design concept $[M_{u,c} > M_{u,b}]$, the maximum value of the collapse multiplier is obtained on the extreme spans when bending moments at these sections reach their ultimate values. In general, it should also be verified that these bending moments shall not be greater than the ultimate moment. Thus, the lower bound of the collapse load, K_s, is given in a more simplified form as

$$K_s = \frac{2(m+1)\, M_{u,b}}{\left[\sum_{i=1}^{n} F_i + \sum_{i=1}^{n-1} F_i \right] H} = \frac{(1+m)(1+n)(1+2n)M_{u,b}}{H(1+2n)^2 V_b} \le K_c \qquad (4.8)$$

However, this simplified procedure cannot be extended for building frames with irregular structural configurations.

4.3.3 STEP-BY-STEP ANALYSIS FOR A SIMPLE FRAME WITH P-M INTERACTION

A step-by-step procedure based on successive applications of the displacement method is briefly presented, where the lateral load (seismic load distributed along the

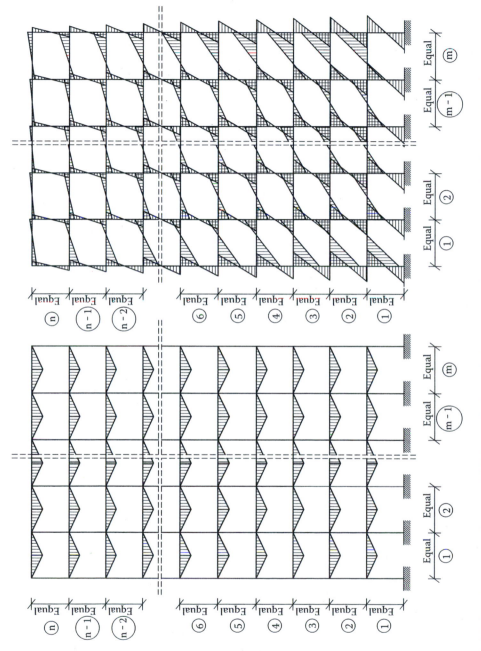

FIGURE 4.6 Bending moment diagrams of RC frames (1) for vertical concentrated loads on beams; (2) distribution of floor shear.

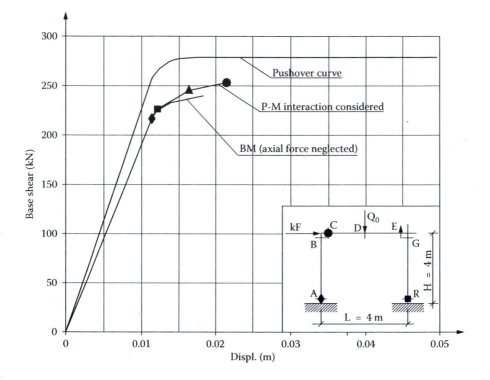

FIGURE 4.7 Force displacement curves by step-by-step analysis (with P-M interaction).

height from base shear) with constant collapse multiplier in each floor is applied until the required number of plastic hinges are formed, leading to collapse. For simplicity, a single story–single bay frame is considered, as shown in Figure 4.7.

Step 1: The frame is characterized by seven sections (A, B, C, D, E, G, R) at which bending moment and axial forces are computed. Three degrees of freedom, namely, θ_C, θ_E, and Δ as rotations at beam-column joints and sway displacement at the top, respectively, are considered. Equilibrium equations, as functions of the degrees of freedom, are given by

$$K \cdot \delta = \beta \tag{4.9}$$

where

$$K = \begin{bmatrix} 4\left(\dfrac{k_b}{L} + \dfrac{k_c}{H}\right) & \dfrac{2k_b}{L} & \dfrac{6k_c}{H^2} \\[3mm] \dfrac{2k_b}{L} & 4\left(\dfrac{k_b}{L} + \dfrac{k_c}{H}\right) & \dfrac{6k_c}{H^2} \\[3mm] \dfrac{6k_c}{H^2} & \dfrac{6k_c}{H^2} & \dfrac{6k_c}{H^3} \end{bmatrix}, \quad \delta = \begin{bmatrix} \theta_C \\[2mm] \theta_E \\[2mm] \Delta \end{bmatrix}, \quad \beta = \begin{bmatrix} -\dfrac{Q_0 L}{8} \\[3mm] \dfrac{Q_0 L}{8} \\[3mm] k_1 F \end{bmatrix} \tag{4.10}$$

where k_b and k_c are stiffness of beam and column elements, respectively. While the vertical load, Q_0, is kept constant, the lateral load, F, is increased by the multiplier k_1. By solving Equation 4.9 with respect to the degrees of freedom, elastic solution for the frame, as a function of the collapse multiplier, is obtained. Bending moment and axial forces at all the sections are given by

$$M_A = -\frac{2k_c}{H}\left[\theta_C + \frac{3}{H}\Delta\right], \quad M_B = \frac{2k_c}{H}\left[2\theta_C + \frac{3}{H}\Delta\right], \quad M_C = -\frac{2k_b}{L}[2\theta_C + \theta_E] - \frac{Q_0 L}{8},$$

$$M_D = -\frac{k_b}{L}[\theta_C - \theta_E] + \frac{Q_0 L}{8}, \quad M_E = \frac{2k_b}{L}[\theta_C + 2\theta_E] - \frac{Q_0 L}{8},$$

$$M_G = +\frac{2k_c}{H}\left[2\theta_E + \frac{3}{H}\Delta\right], \quad M_R = -\frac{2k_c}{H}\left[\theta_E + \frac{3}{H}\Delta\right],$$

(4.11)

$$P_A = P_B = +\frac{6k_b}{L}(\theta_C + \theta_E) + \frac{Q_0}{2}, \quad P_C = P_D = P_E = k_1 F - \frac{6k_b}{H^2}\theta_C - \frac{12k_c}{H^3}\Delta,$$

(4.12)

$$P_G = P_R = -\frac{6k_b}{L}(\theta_C + \theta_E) + \frac{Q_0}{2}$$

By increasing the multiplier, k_1 is obtained as 15.90, while the couple (P, M) in section A reaches the boundary of the P-M domain for columns, resulting in the formation of the first plastic hinge at this section (Figure 4.7). The couples (P, M) at other sections are verified for not reaching the boundaries of their corresponding domains.

Step 2: In the second step, only the lateral load is increased by the multiplier, k_2. With the presence of plastic hinge at section A, the couple (P, M) must belong to P - M domain of the column in the second step. For the equilibrium condition given by Equation 4.9, stiffness matrix and vector β are given by

$$K = \begin{bmatrix} \dfrac{4k_b}{L} + \dfrac{3k_c}{H} & \dfrac{2k_b}{L} & \dfrac{3k_c}{H^2} \\[2mm] \dfrac{2k_b}{L} & 4\left(\dfrac{k_b}{L} + \dfrac{k_c}{H}\right) & \dfrac{6k_c}{H^2} \\[2mm] \dfrac{3k_c}{H^2} & \dfrac{6k_c}{H^2} & \dfrac{15k_c}{H^3} \end{bmatrix}, \quad \beta = \begin{bmatrix} 0 \\ 0 \\ k_2 F \end{bmatrix}$$

(4.13)

Bending moment and axial forces at all section are now given by

$$M_A = 0, \quad M_B = \frac{3k_c}{H}\left[\theta_C + \frac{\Delta}{H}\right], \quad M_C = -\frac{2k_b}{L}[2\theta_C + \theta_E], \quad M_D = -\frac{k_b}{L}[\theta_C - \theta_E]$$

$$M_E = \frac{2k_b}{L}[\theta_C + 2\theta_E], \quad M_G = +\frac{2k_c}{H}\left[2\theta_E + \frac{3}{H}\Delta\right], \quad M_R = -\frac{2k_c}{H}\left[\theta_E + \frac{3}{H}\Delta\right],$$

(4.14)

$$P_A = P_B = +\frac{6k_b}{L}(\theta_C + \theta_E),$$

$$P_C = P_D = P_E = k_2 F - \frac{3k_c}{H^2}\left(\theta_C + \frac{\Delta}{H}\right), \tag{4.15}$$

$$P_G = P_R = -\frac{6k_b}{L}(\theta_C + \theta_E)$$

Net bending moment and axial force at any section can be determined by summarizing the respective equations of Steps 1 and 2. By further increasing the collapse multiplier, k_2 is obtained as 0.5, and the couple (P, M) at section R reaches the boundary of P-M domain of the column, resulting in the formation of the second plastic hinge at this section.

Step 3: Now, a new frame characterized by two plastic hinges at sections A and R is considered. Stiffness matrix and the vector β are given by

$$K = \begin{bmatrix} \dfrac{4k_b}{L} + \dfrac{3k_c}{H} & \dfrac{2k_b}{L} & \dfrac{3k_c}{H^2} \\[2mm] \dfrac{2k_b}{L} & \dfrac{4k_b}{L} + \dfrac{3k_c}{H} & \dfrac{3k_c}{H^2} \\[2mm] \dfrac{3k_c}{H^2} & \dfrac{3k_c}{H^2} & \dfrac{6k_c}{H^3} \end{bmatrix}, \quad \beta = \begin{bmatrix} 0 \\ 0 \\ k_3 F \end{bmatrix}, \tag{4.16}$$

Bending moment and axial forces at all sections are given by

$$M_A = 0, \quad M_B = \frac{3k_c}{H}\left[\theta_C + \frac{\Delta}{H}\right], \quad M_C = -\frac{2k_b}{L}[2\theta_C + \theta_E], \quad M_D = -\frac{k_b}{L}[\theta_C - \theta_E]$$

$$M_E = \frac{2k_b}{L}[\theta_C + 2\theta_E], \quad M_G = \frac{3k_c}{H}\left[\theta_C + \frac{\Delta}{H}\right], \quad M_R = 0, \tag{4.17}$$

$$P_A = P_B = +\frac{6k_b}{L}(\theta_C + \theta_E),$$

$$P_C = P_D = P_E = k_3 F - \frac{3k_c}{H^2}\left(\theta_C + \frac{\Delta}{H}\right), \tag{4.18}$$

$$P_G = P_R = -\frac{6k_b}{L}(\theta_C + \theta_E)$$

Net bending moment and axial force at any section can now be determined by summarizing the respective equations of Steps 1, 2, and 3. By further increasing the collapse multiplier, k_3 is obtained as 1.35, and the couple (P, M) at section E reaches the boundary of the P-M domain of the beam, resulting in the formation of third plastic hinge in the beam at section E.

Step 4: Finally, the frame is characterized by three plastic hinges at sections A, R, and E, formed successively. The stiffness matrix and vector displacements are given by

$$K = \begin{bmatrix} 3\left(\dfrac{k_b}{L} + \dfrac{k_c}{H}\right) & \dfrac{3k_c}{H^2} \\ \dfrac{3k_c}{H^2} & \dfrac{3k_c}{H^3} \end{bmatrix}, \quad \delta = \begin{bmatrix} \theta_C \\ \Delta \end{bmatrix}, \quad \beta = \begin{bmatrix} 0 \\ k_4 F \end{bmatrix} \tag{4.19}$$

By solving, bending moment and axial forces at all sections are given by

$$M_A = 0, \quad M_B = \frac{3k_c}{H}\left[\theta_C + \frac{\Delta}{H}\right],$$

$$M_C = -\frac{3k_c}{L}\theta_C, \quad M_D = -\frac{3k_c}{2L}\theta_C, \quad M_E = 0, \quad M_G = 0, \quad M_R = 0 \tag{4.20}$$

$$P_A = P_B = +\frac{3k_b}{L^2}\theta_C, \quad P_C = P_D = P_E = 0, \quad P_G = P_R = -\frac{3k_b}{L^2}\theta_C \tag{4.21}$$

Net bending moment and axial force at collapse can be obtained by summarizing the respective equations of all the above steps. By further increasing the collapse multiplier, k_4 is obtained as 0.58, and the couple (P, M) at section C reaches the boundary of P-M domain of the beam, resulting in the formation of the final plastic hinge at section C, causing collapse. The total collapse load multiplier is given by the sum of multipliers of each step and is equal to 18.33.

Table 4.1 shows the trace of strain values for concrete and steel at compression and tension, respectively, obtained during the analysis. It can be seen from Table 4.1 that the strain in tensile steel reaches ultimate value, causing the plastic hinges at

TABLE 4.1
Strain Values in Elements Obtained by Step-by-Step Procedure (P-M Interaction)

Step No.	Element	Section	ε_{sc}	ε_{st}	$\varepsilon_{c,max}$
1	column	1	0.0010	0.01000	0.00214
2	column	7	0.0013	0.01000	0.00249
3	beam	5	0.0009	0.01000	0.00213
4	beam	3	0.0013	0.01000	0.00253

TABLE 4.2

Collapse Multiplier, Displacement, and Base Shear Obtained by Step-by-Step Procedure (P-M Interaction)

Step No.	Section	P (kN)	M (kN-m)	Hinge Formation	Collapse Multiplier in Each Stage	Collapse Multiplier	Displ. (m)	Base Shear (kN)
	A	− 68.638	− 253.218	yes				
	B	− 68.638	163.380	no				
	C	115.509	163.380	no				
1	D	115.509	26.103	no	15.900	15.900	0.01152	219.6585
	E	115.509	− 193.673	no				
	G	109.888	193.673	no				
	R	109.888	− 268.364	no				
	A	− 72.262	− 251.861	yes				
	B	− 72.262	169.560	no				
	C	120.872	169.560	no				
2	D	120.872	25.035	no	0.500	16.400	0.012123	226.566
	E	120.872	− 201.989	no				
	G	113.512	201.989	no				
	R	113.512	− 284.451	yes				
	A	− 90.912	− 248.454	yes				
	B	− 90.912	206.860	no				
	C	130.197	206.860	no				
3	D	130.197	25.035	no	1.350	17.750	0.016360	245.2163
	E	130.197	− 239.290	yes				
	G	132.162	239.290	no				
	R	132.162	− 288.856	yes				
	A	− 98.93	− 246.99	yes				
	B	− 98.93	238.91	no				
	C	130.20	238.91	yes				
	D	130.20	41.06	no				
4	E	130.20	− 239.29	yes	0.580	18.330	0.021488	253.229
	G	140.18	239.29	no				
	R	140.18	− 290.30	yes				

critical sections. Collapse is caused by tensile failure as the strain in steel reaches ultimate value before concrete reaches its ultimate strain. Table 4.2 shows the history of collapse multipliers thus obtained along with the displacements and base shear at each step. The strains at critical sections where plastic hinges are formed are verified for their ultimate values. Figure 4.7 shows the force-displacement profile obtained by the force-controlled, step-by-step procedure employed on a single

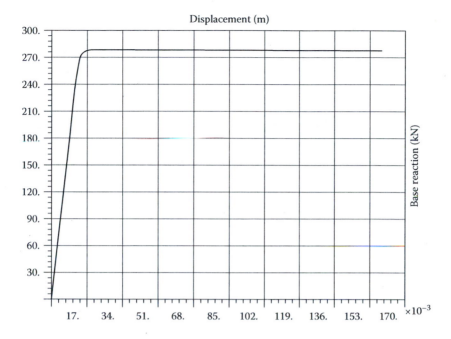

FIGURE 4.8 Pushover curve of single story–single bay RC frame.

story–single bay frame. Traces of plastic hinges tagged in the force-displacement curve can be seen in the figure. Figure 4.8 shows the pushover curve obtained from nonlinear static pushover analysis; while for easy comparison it is also superposed in Figure 4.7. While the P-M interaction is ignored, the final collapse multiplier is 17.35, which marginally underestimates the collapse load in comparison to the case when P-M interaction is considered. Trace of hinges formed during the analysis is shown in Table 4.3. It is worthwhile to note that the history of formation of plastic hinges is different for the two cases, namely, (1) considering axial force and P-M interaction and (2) neglecting axial force, respectively.

4.4   NUMERICAL STUDIES AND DISCUSSIONS

Reinforced concrete building frames with different geometry are analyzed, and bounds of collapse multipliers obtained by employing different procedures are compared. Seven frames are considered for the analytical study: (1) single bay–single story, (2) single bay–double story, (3) single bay–single story with unequal column length, (4) four bay–two story, (5) six bay–three story irregular, (6) six bay–three story regular, and (7) five bay–ten story. All the frames are comprised of (1) 450 mm square RC columns, reinforced with 12#25Φ and lateral ties of 8 mm at 200 c/c (refer to Figure 4.2); (2) 300 × 450 mm RC beam, reinforced with 4#22Φ as tensile and compression steel with shear stirrups of 10 mm at 250 c/c (refer to Figure 4.1); and

TABLE 4.3
Collapse Multiplier, Displacement, and Base Shear Obtained by Step-by-Step Procedure (Neglecting Axial Force)

Step No.	Section	M (kN-m)	Hinge Formation	Collapse Multiplier in Each Stage	Collapse Multiplier	Displ. (m)	Base Shear (kN)
	A	− 249.914	no				
	B	161.124	no				
	C	161.124	no				
1	D	26.104	no	15.699	15.70	0.0114	216.88
	E	− 191.417	no				
	G	191.417	no				
	R	− 265.060	yes				
	A	− 265.060	yes				
	B	170.716	no				
	C	170.716	no				
2	D	27.336	no	0.577	16.28	0.0121	224.85
	E	− 198.544	no				
	G	198.544	no				
	R	− 265.060	yes				
	A	− 265.060	yes				
	B	186.622	no				
	C	186.622	no				
	D	27.336	no				
	E	− 214.450	yes				
3	G	214.450	no	0.576	16.85	0.0139	232.80
	R	− 265.060	yes				
	A	− 265.060	yes				
	B	214.450	no				
	C	214.450	yes				
4	D	107.225	no	0.504	17.35	0.0183	239.76
	E	− 214.450	yes				
	G	214.450	no				
	R	− 265.060	yes				

(3) 125-mm-thick RC slab. M25 mix and high-yield strength deformed bars (Fe 415) are used in the members. All building frames consisting of 4 m bay widths and 4 m story heights are assumed to be located in Zone V (IS 1893, 2002) with soil condition as "medium" type. Seismic weight at each floor is computed using IS code (IS 1893, 2002), and the base shear is distributed along the height of the building. Live load of equivalent magnitude is considered to act at the midspan of the beam, while lateral loads, computed from the base shear, are assumed to act at each floor level.

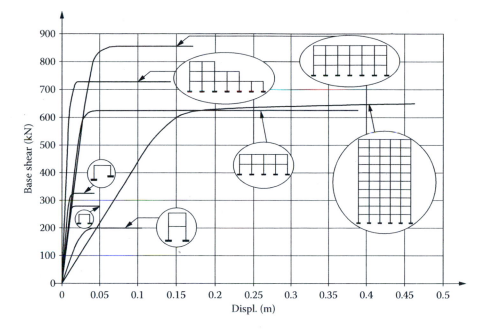

FIGURE 4.9 Pushover curves.

Using the proposed expressions for P-M interaction and moment-rotation, beams and columns are modeled.

Collapse multipliers are assessed by employing above-described procedures, namely, (1) displacement-controlled nonlinear static pushover; (2) upper bound, or kinematic theorem; (3) lower bound, or static theorem; and (4) step-by-step load increment procedure by employing force-controlled method; and the results obtained are compared. Nonlinear characteristics of beam and column hinges are assigned to the structural elements of the building frames and performance levels, namely, (1) immediate occupancy (IO), (2) life safety (LS), and (3) collapse prevention (CP), are tagged to the respective moment-rotation curves during the pushover analysis. Displacement-controlled pushover analysis is performed for the preset target displacement of about 4% of the height of the building to trace the formation of plastic hinges. Pushover curves obtained are plotted for different types of building frames considered and shown in Figure 4.9. Base force corresponding to the step at which requisite numbers of plastic hinges are formed to ensure a collapse mechanism is traced and tabulated. Collapse multiplier is obtained as the ratio of base shears at collapse and design base shear recommended by the code for safe seismic design (IS 1893, 2002). Collapse multipliers obtained using (1) kinematic procedure (Equation 4.4), (2) modified static procedure (using Equation 4.8), and (3) mathematical programming tool (LINGO) (Raphel, Marak, and Truszcynski 2002; Sforza 2002) are also shown in the table. During limit analysis procedures, P-M interactions of the structural elements are ignored.

To trace the path of formation of plastic hinges, a step-by-step, force-controlled procedure is also employed on a single bay–single story frame, with and without considering P-M interaction. Plastic hinges obtained at each step are traced and the corresponding displacement and collapse multipliers are recorded. Tables 4.2 and 4.3 show the displacement, base shear, and collapse multipliers obtained at various steps for both cases, (1) considering P-M interaction and (2) neglecting axial force, respectively. It can be seen from the tables that the collapse multipliers obtained from the force-controlled, step-by-step procedure are 18.33 (by considering P-M interaction) and 17.35 (by neglecting axial force). A ready comparison of displacement-controlled pushover cannot be made with the force-controlled method, employed on a single bay–single story frame. However, base shear obtained from pushover analysis at Step 6 (refer to Table 4.4), where four plastic hinges are formed (same as the case of the step-by-step, force-controlled procedure), becomes comparable and the collapse multiplier determined from pushover is 20. Table 4.4 also shows that the collapse load at this stage determined from pushover is capable of *pushing* the rooftop of the frame by about 14 mm as compared to that of about 21 mm and 18 mm obtained from the force-controlled methods (refer to Tables 4.2 and 4.3); hence, the comparison is made by considering the force level causing the same number of plastic hinges.

Figure 4.10 shows the comparison of collapse multipliers obtained for different types of buildings considered in the study, while Table 4.4 shows the comparison obtained by employing different procedures. By comparing these multipliers, it can be seen that for a single bay–single story frame, plastic theorems underestimate the true collapse load in comparison with pushover, since they do not account for reserve capacities of structural members that can be reflected in the analysis by considering P-M interaction. For multibay-multistory frames, pushover multipliers closely agree with that of the kinematic theorem; also in the case of frames with irregular structural configurations, pushover multipliers are in close agreement with kinematic theorem only for increased bay and story numbers. This intuits employing kinematic theorem as an approximate method for the preliminary estimate of collapse loads, which should be subsequently verified by pushover analysis, however. Limitations imposed by mathematical programming tools can be seen for the absence of results for higher story frames (for example, a ten-story frame), whereas no such limitations are imposed by pushover analysis. Force-controlled, step-by-step analysis is capable of estimating the collapse multiplier in close agreement with pushover and is a better estimate compared with limit theorems. This may be due to the fact that the former method accounts for redistribution of moment-carrying capacity of plastic hinges at critical sections, which is an indirect contribution from P-M interaction. While axial force is neglected, this procedure results in the same value as that of the limit theorems, since the hypothesis becomes the same in both the cases.

TABLE 4.4
Collapse Multipliers Obtained from Different Procedures

Structure	Collapse Load Multiplier					Pushover Analysis Results											
	k_k (Eq 4.4)	k_s (LINGO)	k_k (Eq 4.8)	$k_{pushover}$	V_B (kN)	Step No.	Displ. (mm)	V_B (kN)	HINGE FORMATION								No. plastic hinges
									A to B	B to IO	IO to LS	LS to CP	CP to C	C to D	D to E	Total	
	17.35	17.35	15.52	20.00	13.82	6	14	276.238	2	1	1	0	0	0	2	6	4
	6.97	6.97	6.47	7.17	27.63	6	40	198.066	6	1	1	0	0	0	4	12	6
	20.59	20.59	–	23.79	13.58	4	14	323.195	2	1	0	0	0	0	3	6	4
	6.63	5.95	4.49	6.28	99.59	20	388	625.633	14	2	1	0	0	0	19	36	22
	5.01	4.52	–	4.50	149.38	6	13	671.861	20	1	1	0	0	0	32	54	34
	4.21	3.68	2.50	3.86	221.32	9	75	855.094	40	0	0	0	0	0	38	78	38
	2.60	Limitation	1.31	2.32	281.76	11	600	653.12	107	7	5	0	0	0	101	220	113

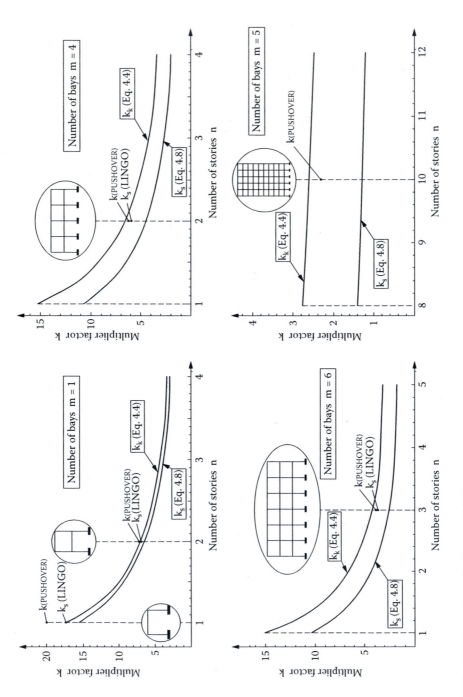

FIGURE 4.10 Comparison of collapse multipliers.

4.5   CONCLUSIONS

Although different procedures exist to estimate collapse loads of building frames under seismic action, a relatively new procedure, pushover analysis, is compared, showing its suitability for the subject of discussion. Based on the numerical studies conducted, it can be seen that for single-story building frames, plastic theorems underestimate the true collapse load in comparison with pushover because they do not account for reserve capacities of structural elements that can be reflected in the analysis by considering P-M interaction. For multistory-multibay frames, design base shear estimated by pushover closely agrees with the kinematic theorem, making it an appropriate method for the preliminary estimate of collapse load in such cases. Force-controlled, step-by-step analysis is capable of estimating the collapse load in close agreement to pushover (closer than limit theorems) because it accounts for P-M interaction. But still the difference may be due to the fact that the former procedure does not account for redistribution of moments at critical sections where plastic hinges are formed. Also, this procedure is computationally expensive and cumbersome in comparison with nonlinear static pushover.

Under the increasing necessity of seismic evaluation of existing RC buildings, displacement-controlled pushover analysis is certainly seen as an appropriate and reasonably accurate tool. This study shall help the designers and researchers to use displacement-controlled pushover analysis with improved confidence as their results of different examples are compared with other similar methods used to assess the collapse loads. The results obtained are influenced by input parameters, P-M interaction in particular. It is therefore emphasized to use axial force–bending moment yield interaction accounting for nonlinear characteristics of constitutive materials. Though the results obtained by employing plastic theorems on the limited examples are not new, the study quantifies these values through illustrated examples and their comparison with those obtained using pushover analysis; this is a relatively new attempt made through this study. With the presented mathematical modeling and proposed expressions for the said input parameters to nonlinear static pushover analysis, it is believed that designers and researchers will use pushover analysis more commonly in the future with improved confidence and accuracy.

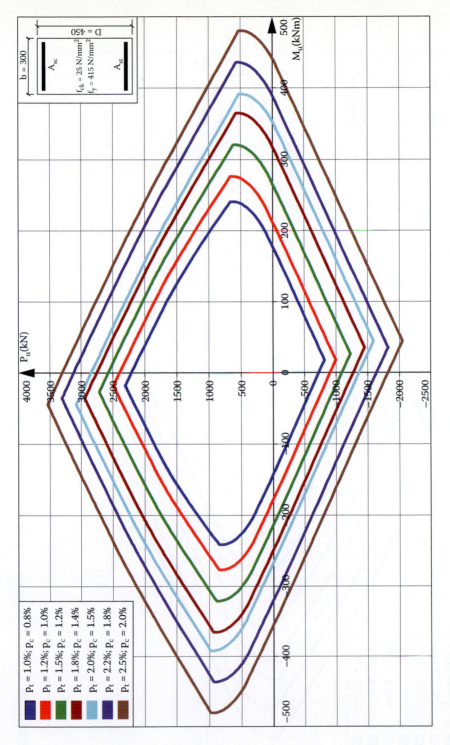

COLOR FIGURE 1.9 P-M interaction curves for RC section 300×450 ($f_{ck} = 25 \text{ N/mm}^2$, $f_y = 415 \text{ N/mm}^2$).

Legend:
- $P_t = 1.0\%$; $P_c = 0.8\%$
- $P_t = 1.2\%$; $P_c = 1.0\%$
- $P_t = 1.5\%$; $P_c = 1.2\%$
- $P_t = 1.8\%$; $P_c = 1.4\%$
- $P_t = 2.0\%$; $P_c = 1.5\%$
- $P_t = 2.2\%$; $P_c = 1.8\%$
- $P_t = 2.5\%$; $P_c = 2.0\%$

Section details: $b = 300$, $D = 450$, A_{sc}, A_{st}, $f_{ck} = 25 \text{ N/mm}^2$, $f_y = 415 \text{ N/mm}^2$.

Axes: P_u (kN), M_u (kNm).

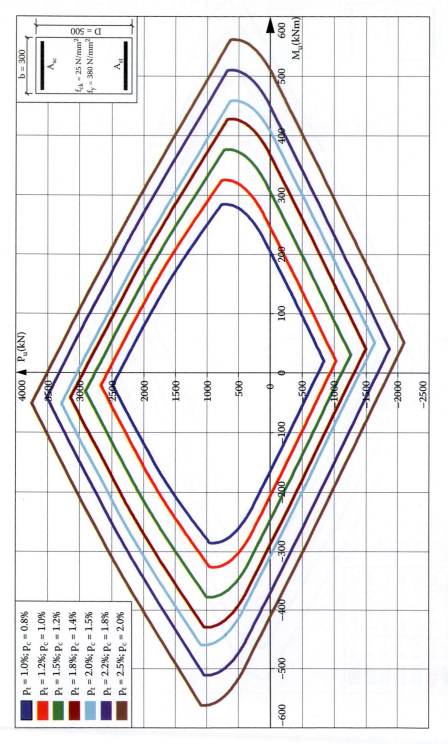

COLOR FIGURE 1.10 P-M interaction curves for RC section 300 × 500 (f_{ck} = 25 N/mm², f_y = 380 N/mm²).

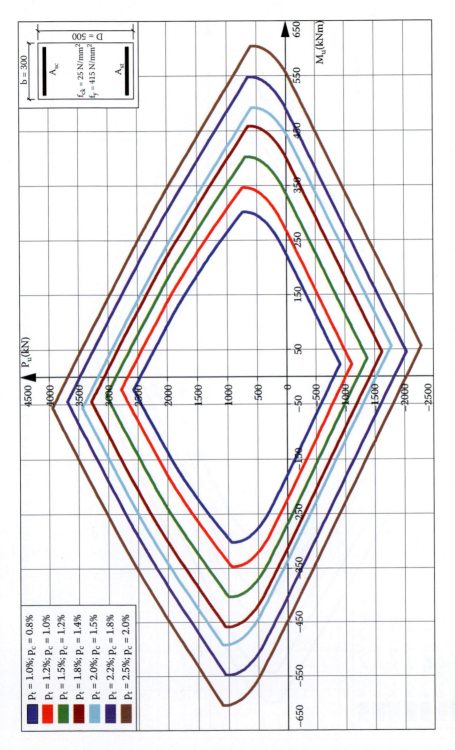

COLOR FIGURE 1.11 P-M interaction curves for RC section 300×500 ($f_{ck} = 25$ N/mm², $f_y = 415$ N/mm²).

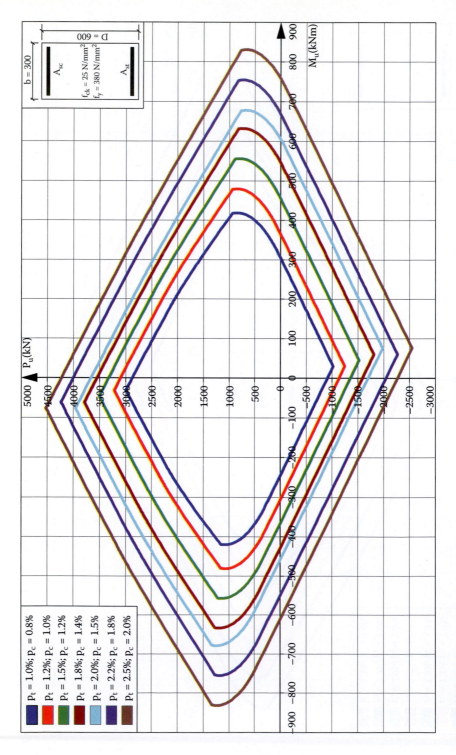

COLOR FIGURE 1.12 P-M interaction curves for RC section 300×600 ($f_{ck} = 25$ N/mm^2, $f_y = 380$ N/mm^2).

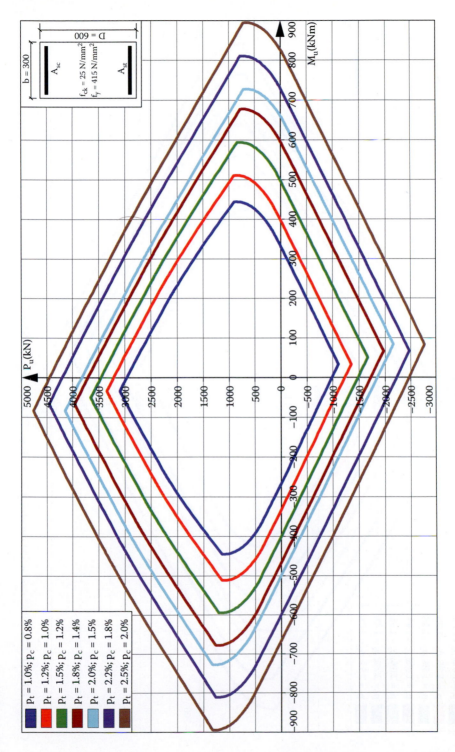

COLOR FIGURE 1.13 P-M interaction curves for RC section 300 × 600 (f_{ck} = 25 N/mm², f_y = 415 N/mm²).

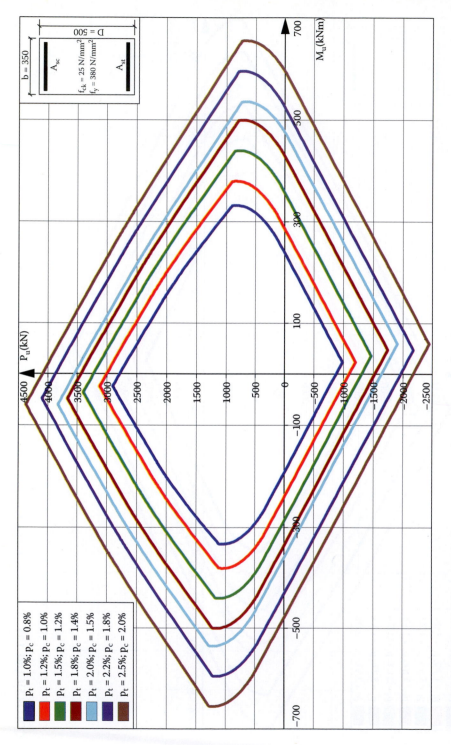

COLOR FIGURE 1.14 P-M interaction curves for RC section 350 × 500 (f_{ck} = 25 N/mm², f_y = 380 N/mm²).

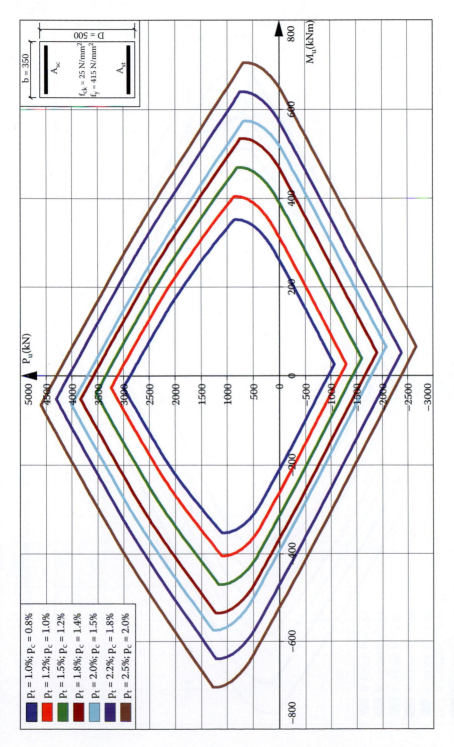

COLOR FIGURE 1.15 P-M interaction curves for RC section 350×500 ($f_{ck} = 25$ N/mm^2, $f_y = 415$ N/mm^2).

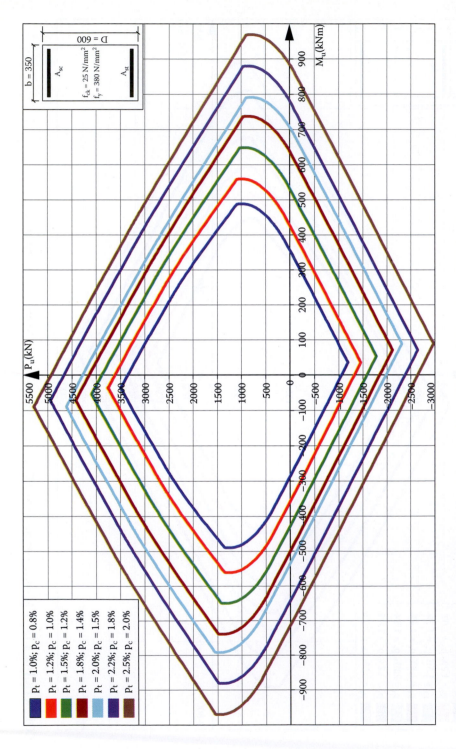

COLOR FIGURE 1.16 P-M interaction curves for RC section 350 × 600 ($f_{ck} = 25$ N/mm², $f_y = 380$ N/mm²).

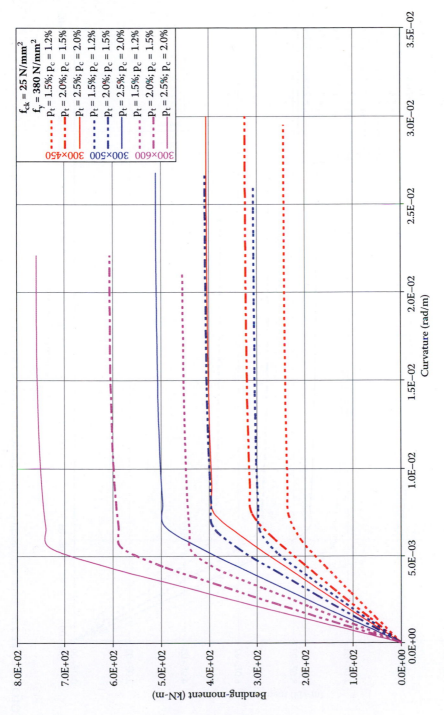

COLOR FIGURE 2.8 Bending moment-curvature for RC sections 300 mm wide (f_{ck} = 25 N/mm², f_y = 380 N/mm²).

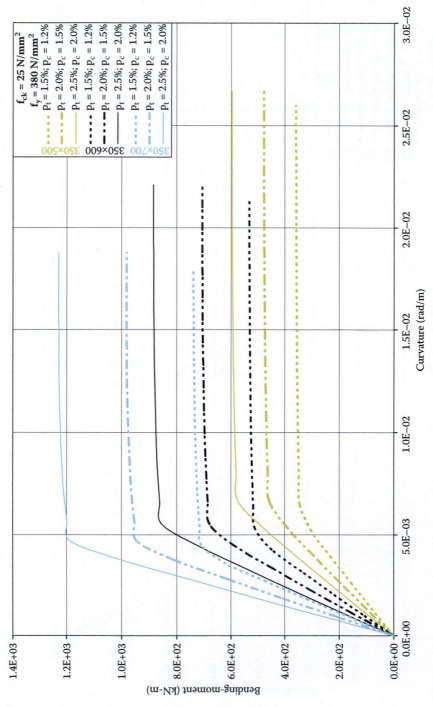

COLOR FIGURE 2.9 Bending moment-curvature for RC sections 350 mm wide ($f_{ck} = 25$ N/mm², $f_y = 380$ N/mm²).

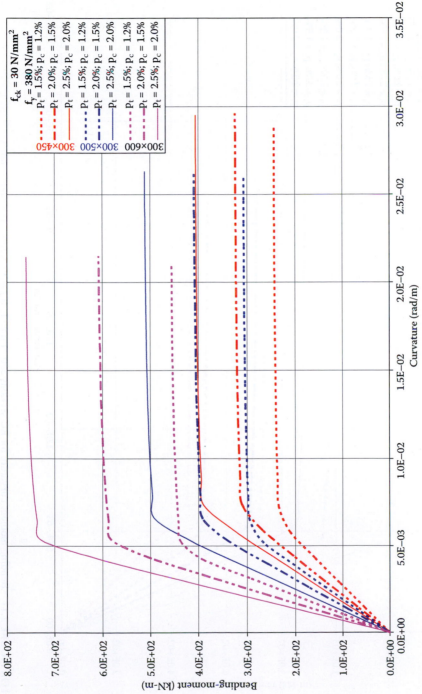

COLOR FIGURE 2.10 Bending moment-curvature for RC sections 300 mm wide ($f_{ck} = 30$ N/mm², $f_y = 380$ N/mm²).

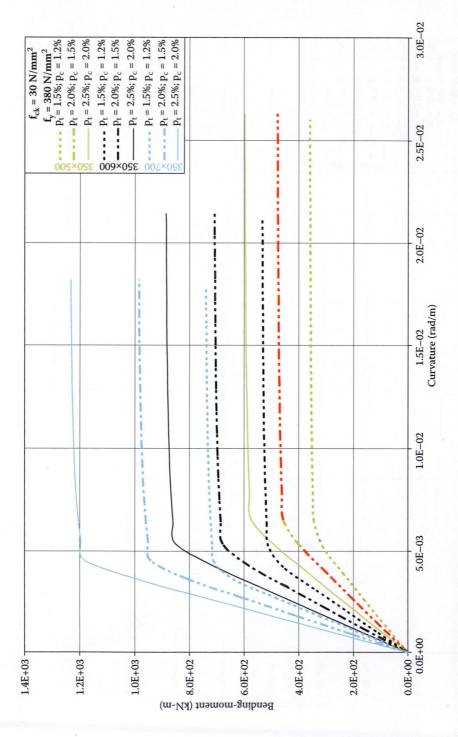

COLOR FIGURE 2.11 Bending moment-curvature for RC sections 350 mm wide (f_{ck} = 30 N/mm², f_y = 380 N/mm²).

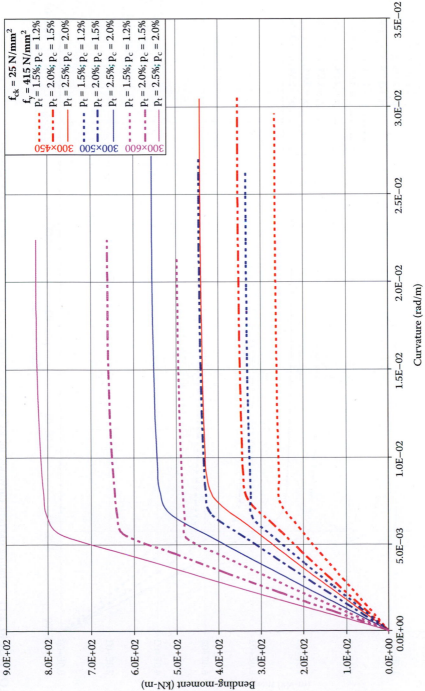

COLOR FIGURE 2.12 Bending moment-curvature for RC sections 300 mm wide (f_{ck} = 25 N/mm², f_y = 415 N/mm²).

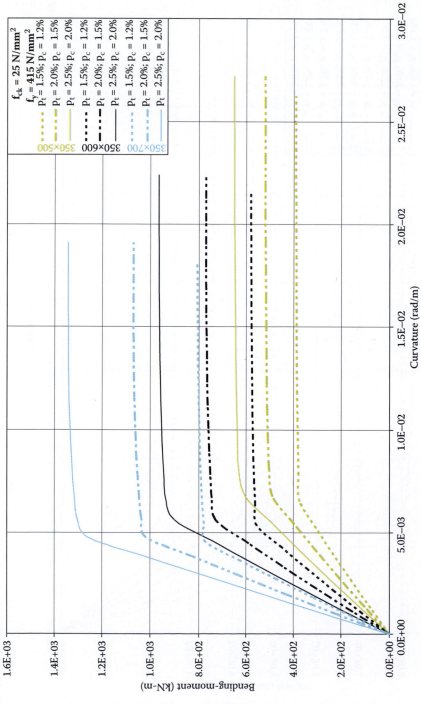

COLOR FIGURE 2.13 Bending moment-curvature for RC sections 350 mm wide (f_{ck} = 25 N/mm², f_y = 415 N/mm²).

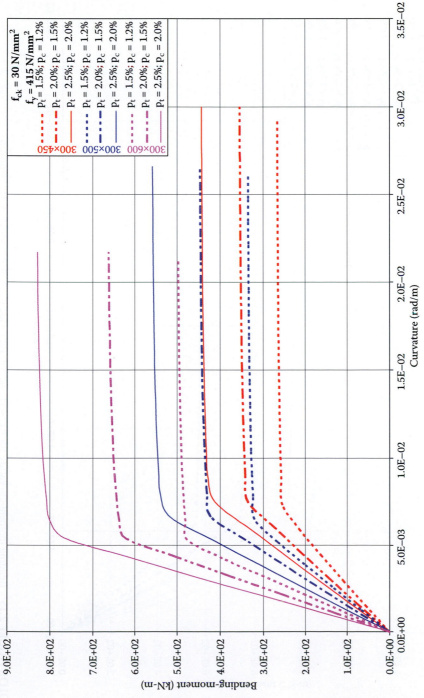

COLOR FIGURE 2.14 Bending moment-curvature for RC sections 300 mm wide ($f_{ck} = 30$ N/mm², $f_y = 415$ N/mm²).

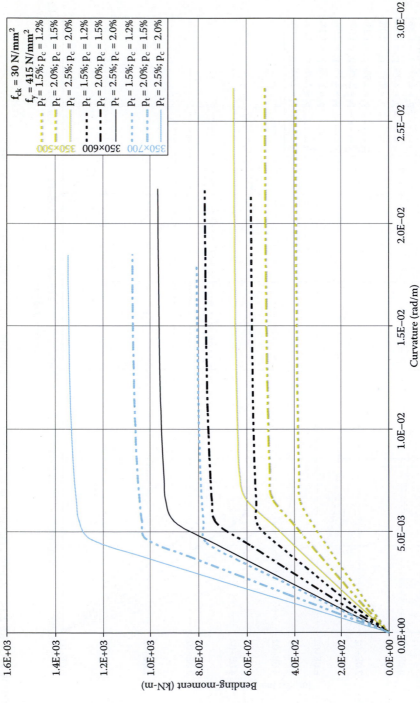

COLOR FIGURE 2.15 Bending moment-curvature for RC sections 350 mm wide ($f_{ck} = 30$ N/mm², $f_y = 415$ N/mm²).

5 Flow Rule Verification for P-M Interaction Domains

5.1 SUMMARY

A detailed analytical modeling of P-M yield interaction is presented in Chapter 1, defining the limit boundaries with six subdomains based on Eurocode currently in prevalence. In this chapter, the developed P-M interaction domains are verified for plastic flow-rule in two main sections: (1) tension failure resulting in yielding of steel and (2) compression failure resulting in crushing of concrete. The conventional limit P-M domain is described according to Eurocode currently in prevalence as long as the plastic strain increment becomes nearly normal to the yield domain over the part of bending response, in the presence of axial force. The flow rule verifies for a close agreement in all subdomains of tension failure, while it does not qualify in a few of the subdomains of crushing failure. The mathematically developed P-M interaction model is thus capable of identifying the damage mechanism of different subdomains in RC sections, in a closer agreement for tension failure subdomains, in particular; damage identification is made on the basis of strain profile of concrete and reinforcing steel.

5.2 INTRODUCTION

Earlier studies conducted by researchers (e.g., Abu-Lebdeh and Voyiadjis 1993; Park and Kim 2003) emphasized a prerequisite of material response behavior to a variety of loads to successfully forecast their behavior under conditions leading to damage. Through knowledge of P-M yield domains, the structural designer is enabled to assess the type of failure caused to the member, either tensile or compressive (Chandrasekaran et al. 2008a). Khan, Al-Gadhib, and Baluch (2007) used two parameters to define the effective compliance nature of elastic-damage model of high-strength concrete under multiaxial loading; these two parameters account for different behavior of concrete in tension and compression. They emphasized that the study of concrete behavior under P-M interaction is necessary to trace the strain softening effect, in particular. In a reinforced concrete section, reinforcement behaves as an elastic-plastic spring because of which an RC beam section developed horizontal cracks under three-point loading (Sumarec, Sekulovic, and Krajcinovic 2003). It is a well-understood fact that RC members inherit the flexibility of changing, within certain limits, the ultimate moment as the designer pleases, without undergoing a major change in the overall dimensions of the cross-section. This initiated a recent practice among structural designers to adjust the area of tensile steel for achieving the distribution of ultimate moment to be the same as that of the elastic moment for the factored load. However, under seismic loads, codes insist that the structures should be designed to resist earthquakes in a quantifiable manner imposed with desired possible damage (e.g., Ganzerli, Pantelides, and Reaveley 2000). Therefore, damage models that quantify severity of repeated plastic cycling through energy dissipation are simple tools that can be used for safe seismic design. The strain equivalence hypothesis used by the researchers (e.g., Hsieh, Ting, and Chen 1982) that equates strain in effective (undamaged) and damaged configurations is adopted for deriving the constitutive equations in the present study.

The proposed P-M yield interaction shown in Chapter 1 is a conventional limit domain with strain limits prescribed by Eurocode; hence, it does not fulfill the complete mechanical meaning. Since the entire cross-section is not under limit stress, the proposed limit domain is different from the one valid for homogeneous materials like steel. Further, equilibrium states inside the P-M boundary are not fully in *elastic state* since loading and unloading for composite materials, such as reinforced concrete does not follow the same path. Also, the plastic strain increments evaluated for limit stress states belonging to P-M boundary are not truly and completely plastic increments because part of the section remains elastic. The above-mentioned arguments are addressed in this chapter with a main focus to verify the plastic flow rule in the developed P-M interaction domains. To examine this objective closely, P-M domains are reclassified broadly as (1) tensile failure resulting in yielding of steel, which is now subdivided into *five subdomains*, and (2) compression failure resulting in crushing of concrete, which is now further subdivided into *five subdomains*, making the total number of subdomains *ten* instead of *six* as seen in Chapter 1. For closer examination of plastic flow rule, this reclassification becomes inevitable.

5.3 MATHEMATICAL DEVELOPMENT

The domain 2a discussed in Chapter 1 is now subdivided into two, namely, $2_a^{(1)}$ and $2_a^{(2)}$; domain 2b is subdivided into two, namely, $2_b^{(1)}$ and $2_b^{(2)}$ in the tensile failure zone. In the compression failure zone, domain 6 presented in Chapter 1 is now subdivided into two, namely, 6a and 6b. All other subdomains proposed in Chapter 1 remain the same. Figure 5.1 shows the typical P-M limit domain consisting of ten subdomains as discussed below. Only the upper boundary curves will be examined to see which one-to-one M = M(P) relationship exists; the lower boundary can be readily examined using the similar procedure and hence is not presented. Figure 5.2 shows the strain level in steel and concrete for subdomains 1 to $2_b^{(2)}$ in which collapse is caused by yielding of steel, and Figure 5.3 shows strain levels for subdomains 3 to 6b in which collapse is caused by crushing of concrete.

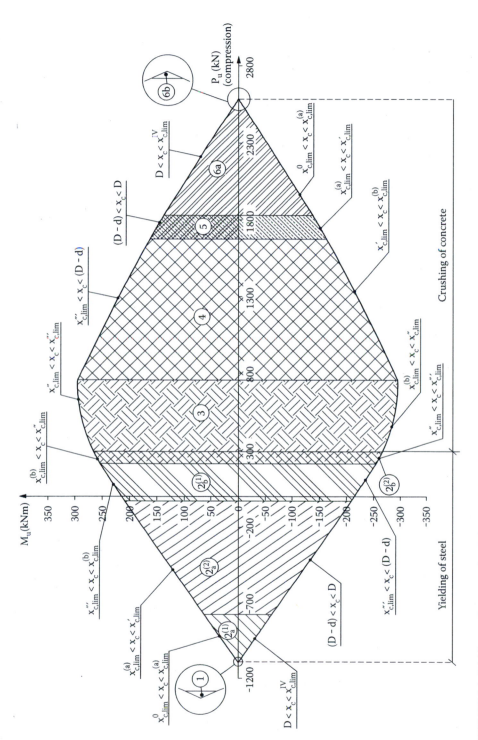

FIGURE 5.1 P-M interaction curve for different subdomains.

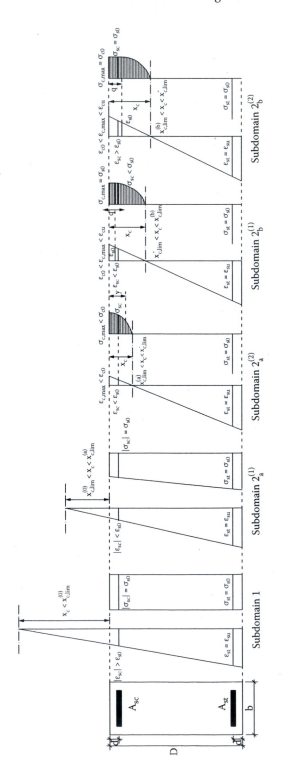

FIGURE 5.2 Collapse caused by yielding of steel.

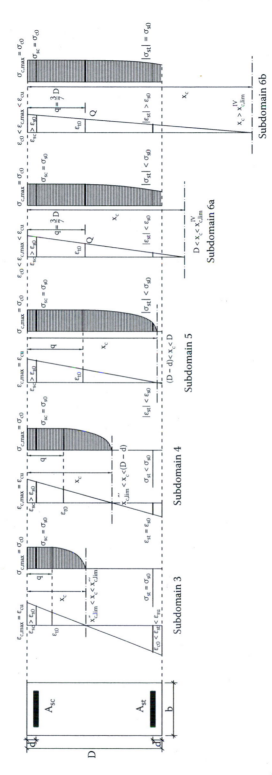

FIGURE 5.3 Collapse caused by crushing of concrete.

5.3.1 Subdomains 1 to $2_b^{(2)}$: Collapse Caused by Yielding of Steel

In the subdomains from 1 to $2_b^{(2)}$, strain in tensile steel reaches its ultimate limit, and the corresponding stress reaches the design ultimate stress; strain in compressive steel is given by

$$\varepsilon_{sc} = \varepsilon_{su}\left(\frac{x_c - d}{D - x_c - d}\right) \tag{5.1}$$

Strain in any generic compression fiber of concrete located at a distance y measure from the extreme compression fiber of concrete is given by

$$\varepsilon_c(y) = \frac{\varepsilon_{su}(x_c - y)}{D - x_c - d}, \quad \varepsilon_{c,max} = \frac{\varepsilon_{su}x_c}{D - x_c - d} \tag{5.2}$$

where $\varepsilon_{c,max}$ is the maximum strain in concrete.

In subdomain 1, neglecting the tensile stress in concrete in the equilibrium equations, the position of the neutral axis lies in the range $]-\infty, x_{c,lim}^{(0)}]$. $x_{c,lim}^{(0)}$ is the limit position of neutral axis between two subdomains 1 and $2_a^{(1)}$ for strain in compression steel reaching its elastic limit (refer to Figure 5.2, subdomain 1). It is important to note that the neutral axis positions are chosen only for detecting the characteristics of the P-M boundary; please note that the succeeding states do not belong to the same loading path for the chosen cross-section. This limit position is given by

$$x_{c,lim}^{(0)} = \frac{d(\varepsilon_{su} + \varepsilon_{s0}) - D\varepsilon_{s0}}{(\varepsilon_{su} - \varepsilon_{s0})} \tag{5.3}$$

In subdomain 1, for strain conditions $\varepsilon_{s0} < \varepsilon_{sc} > \varepsilon_{su}$ and $\sigma_{sc} = \sigma_{s0}$, ultimate axial force and bending moment are given by

$$\begin{cases} P_u = \sigma_{s0}\, b(D-d)(p_c - p_t) \\ M_u = P_u b\left(\dfrac{D}{2} - d\right) \end{cases} \quad \forall x_c \in]-\infty, x_{c,lim}^{(0)}] \tag{5.4}$$

$$p_c = \frac{A_{sc}}{b(D-d)}, \quad p_t = \frac{A_{st}}{b(D-d)} \tag{5.5}$$

where p_t, p_c are percentage of tensile and compression reinforcements, respectively. It may be noted from the above equations that the ultimate axial force and bending moment are independent of the position of neutral axis; bending moment varies linearly with axial force.

In subdomain $2_a^{(1)}$, neglecting the tensile stress in concrete in the equilibrium equations, the position of the neutral axis lies in the range $[x_{c,lim}^{(0)}, x_{c,lim}^{(a)} = 0]$, where $x_{c,lim}^{(a)}$ is the limit position of the neutral axis between two subdomains $2_a^{(1)}$ and $2_a^{(2)}$ for strain in compression steel less than elastic limit (refer to Figure 5.2, subdomain $2_a^{(1)}$). In subdomain $2_a^{(1)}$, for the strain conditions $\varepsilon_{sc} \le \varepsilon_{s0}$, $\sigma_{sc} = E_s \varepsilon_{sc}$, ultimate axial force and bending moment are given by

$$\begin{cases} P_u = b(D-d)\left[p_c E_s \varepsilon_{su} \left(\dfrac{x_c - d}{D - x_c - d} \right) - \sigma_{s0} p_t \right] \\[4mm] M_u = \dfrac{1}{2} b(D-d)(D-2d)\left[p_c E_s \varepsilon_{su} \left(\dfrac{x_c - d}{D - x_c - d} \right) + \sigma_{s0} p_t \right] \end{cases} \quad \forall x_c \in \left[x_{c,lim}^{(0)}, x_{c,lim}^{(a)} = 0 \right]$$

(5.6)

Depth of neutral axis can be deduced as

$$x_c = \frac{(D-d)[P_u + b(E_s p_c \varepsilon_{su} d - \sigma_{s0} p_t (D-d))]}{P_u + b(D-d)(E_s p_c \varepsilon_{su} + p_t \sigma_{s0})}$$

(5.7)

Further, by substituting in Equation 5.6, we get

$$M_u = [P_u + 2b(D-d) p_t \sigma_{s0}]\left(\frac{D}{2} - d \right)$$

(5.8)

Besides, for depth of neutral axis reaching zero, ultimate axial force and bending moment are given by

$$\begin{cases} P_u = -b[d p_c E_s \varepsilon_{su} + (D-d) p_t \sigma_{s0}] \\[4mm] M_u = b\left(\dfrac{D}{2} - d \right)[(D-d) p_t \sigma_{s0} - d p_c E_s \varepsilon_{su}] \end{cases} \quad \text{for } x_c = x_{c,lim}^{(a)} = 0 \quad (5.9)$$

In subdomain $2_a^{(2)}$, stress in any generic compression fiber of concrete is given by

$$\sigma_c(\varepsilon_c(y)) = -\frac{\sigma_{c0}}{\varepsilon_{c0}^2}\varepsilon_c^2 + \frac{2\sigma_{c0}}{\varepsilon_{c0}}\varepsilon_c = \frac{(x_c - y)[2\varepsilon_{c0}(D - x_c - d) + \varepsilon_{su}(y - x_c)]\sigma_{c0}\varepsilon_{su}}{\varepsilon_{c0}^2(x_c + d - D)^2}$$

(5.10)

The position of the neutral axis lies in the range $[x_{c,lim}^{(a)} = 0, x'_{c,lim}]$, where $x'_{c,lim}$ is the limit position of neutral axis between the subdomains $2_a^{(2)}$ and $2_b^{(1)}$ for maximum strain in concrete approaching elastic limit (refer to Figure 5.2, subdomain $2_a^{(2)}$). This limit position is given by

$$x'_{c,lim} = \left(\frac{\varepsilon_{c0}}{\varepsilon_{c0} + \varepsilon_{su}}\right)(D - d) = \left(\frac{0.002}{0.002 + 0.01}\right)(D - d)$$

$$= 0.167(D - d) \qquad \text{for } \varepsilon_{c,max} = \varepsilon_{c0}; \varepsilon_{st} = \varepsilon_{su} \tag{5.11}$$

Ultimate axial force and bending moment in subdomain $2_a^{(2)}$ are given by

$$\begin{cases} P_u = b\left[\int_0^{x_c} b\sigma_c(\varepsilon_c(y))dy + (D - d)(\sigma_{sc}p_c - \sigma_{s0}p_t)\right] \\ \\ M_u = b\left[\int_0^{x_c} b\sigma_c(\varepsilon_c(y))\left(\frac{D}{2} - y\right)dy + (D - d)(\sigma_{sc}p_c + \sigma_{s0}p_t)\left(\frac{D}{2} - d\right)\right] \end{cases} \quad x_{c,lim}^{(a)} \le x_c \le x'_{c,lim} \tag{5.12}$$

It may be noted that the stresses in concrete and compression steel are less than their elastic limits. The ultimate axial force expression given by Equation 5.12 can be rewritten as

$$A_0 + A_1 x_c + A_2 x_c^2 + A_3 x_c^3 = 0 \tag{5.13}$$

where the constants, $A_{i=0,1,2,3}$ are given by the following relationships:

$$A_0 = -(d - D)^2 [P_u + bdE_s p_c \varepsilon_{su} + b(D - d)p_t \sigma_{s0}]$$

$$A_1 = (D - d)[2P_u + bDE_s p_c \varepsilon_{su} + 2b(D - d)p_t \sigma_{s0}]$$

$$A_2 = \frac{-P_u \varepsilon_{c0} + b(d - D)[\varepsilon_{su}(E_s p_c \varepsilon_{s0} - \sigma_{c0}) + p_t \varepsilon_{c0}\sigma_{s0}]}{\varepsilon_{c0}} \tag{5.14}$$

$$A_3 = \frac{-b\varepsilon_{su}(3\varepsilon_{c0} + \varepsilon_{su})\sigma_{c0}}{3\varepsilon_{c0}^2}$$

By solving Equation 5.13, the depth of the neutral axis can be derived as a function of axial force and properties of the cross-section, as given by

$$x_{c1}(P_u) = \frac{1}{6A_3}\left[-2A_2 + \frac{2.5198\left(A_2^2 - 3A_1A_3\right)}{\lambda} + 1.5874\lambda\right]$$

$$x_{c2}(P_u) = \frac{1}{12A_3}\left[-4A_2 - \frac{(2.5198 + 4.3645i)\left(A_2^2 - 3A_1A_3\right)}{\lambda} - (1.5874 - 2.7495i)\lambda\right]$$

$$x_{c3}(P_u) = \frac{1}{12A_3}\left[-4A_2 - \frac{(2.5198 - 4.3645i)\left(A_2^2 - 3A_1A_3\right)}{\lambda} - (1.5874 + 2.7495i)\lambda\right]$$

$$(5.15)$$

where

$$\lambda = \left[\sqrt{-4\left(A_2^2 - 3A_1A_3\right)^3 + \left(2A_2^3 - 9A_1A_2A_3 + 27A_3^2A_0\right)^2} - 2A_2^3 + 9A_1A_2A_3 - 27A_3^2A_0\right]^{1/3}$$

$$(5.16)$$

Out of the above, only one root (x_{c3}) is in close agreement with the numerical solution obtained; by substituting x_{c3} in moment expression of Equation 5.12, we get

$$M_u = B_0 + \left(\frac{1}{D - x_{c3}(P_u) - d}\right)\left[B_1(x_{c3}(P_u) - d) + \frac{B_2 x_{c3}^2(P_u) + B_3 x_{c3}^3(P_u) + B_4 x_{c3}^4(P_u)}{(D - x_{c3}(P_u) - d)}\right]$$

$$(5.17)$$

where the constants, $B_{i=0,1,2,3,4}$ are given by

$$B_0 = \frac{bp_t\sigma_{s0}}{2}(2d^2 - 3dD + D^2)$$

$$B_1 = \frac{1}{2}b(2d^2 - 3dD + D^2)E_sp_c\varepsilon_{su}$$

$$B_2 = \frac{bD(D - d)\varepsilon_{su}\sigma_{c0}}{2\varepsilon_{c0}}$$

$$B_3 = \frac{b\varepsilon_{su}\sigma_{c0}}{6\varepsilon_{c0}^2}[2d\varepsilon_{c0} - D(5\varepsilon_{c0} + \varepsilon_{su})]$$

$$B_4 = \frac{b\varepsilon_{su}}{12\varepsilon_{c0}^2}(4\varepsilon_{c0} + \varepsilon_{su})\sigma_{c0}$$

$$(5.18)$$

In subdomain $2_b^{(1)}$, maximum strain in concrete is greater than elastic limit but still less than its ultimate limit; this causes plasticization of a small zone near the extreme compression fiber. This zone is termed as the *plastic kernel* of concrete, whose depth is given by the following equation:

$$q = x_c - \frac{\varepsilon_{c0}}{\varepsilon_{su}}(D - x_c - d) \tag{5.19}$$

The position of the neutral axis lies in the range $[x'_{c,lim}, x^{(b)}_{c,lim}]$, where $x^{(b)}_{c,lim}$ is the limit position of neutral axis between subdomains $2_b^{(1)}$ and $2_b^{(2)}$ for strain in compression steel approaches elastic limit (refer to Figure 5.2, subdomain $2_b^{(1)}$). This limit position is given by

$$x^{(b)}_{c,lim} = \left(\frac{\varepsilon_{s0}}{\varepsilon_{s0} + \varepsilon_{su}} \right)(D - 2d) + d \quad \text{for } \varepsilon_{sc} = \varepsilon_{s0}; \quad \varepsilon_{st} = \varepsilon_{su} \tag{5.20}$$

Expressions for ultimate axial force and bending moment in subdomain $2_b^{(1)}$ are given by

$$\begin{cases} P_u = b\left[\int_q^{x_c} \sigma_c(\varepsilon_c(y))dy + q\sigma_{c0} + (D-d)(\sigma_{sc}P_c - \sigma_{s0}P_t) \right] \\[2em] M_u = b\left[\int_q^{x_c} \sigma_c(\varepsilon_c(y))\left(\frac{D}{2} - y \right)dy + \frac{q\sigma_{c0}}{2}(D-q) + (D-d)(\sigma_{sc}P_c + \sigma_{s0}P_t)\left(\frac{D}{2} - d \right) \right] \end{cases}$$

$$x'_{c,lim} \le x_c \le x^{(b)}_{c,lim}$$
$$\tag{5.21}$$

Expression for axial force, presented in the above equation can be rewritten as

$$C_0 + C_1 x_c + C_2 x_c^2 = 0 \tag{5.22}$$

where the constants, $C_{i=0,1,2}$ are given by the following relationship:

$$C_0 = \frac{(D-d)\left[3P_u\varepsilon_{su} + b\left(3dE_sP_c\varepsilon_{su}^2 + (D-d)(3p_t\varepsilon_{su}\sigma_{s0} + \varepsilon_{c0}\sigma_{c0}) \right) \right]}{3\varepsilon_{su}}$$

$$C_1 = -\frac{3P_u\varepsilon_{su} + b(d-D)[2\varepsilon_{c0}\sigma_{c0} + 3\varepsilon_{su}(E_sP_c\varepsilon_{su} + \sigma_{c0} + p_t\sigma_{s0})]}{3\varepsilon_{su}} \tag{5.23}$$

$$C_2 = \frac{b\sigma_{c0}}{3}\left(3 + \frac{\varepsilon_{c0}}{\varepsilon_{su}} \right)$$

By solving, the depth of the neutral axis can be derived as a function of axial force and properties of the cross-section as given below:

$$x_c = -\frac{C_1 + \sqrt{C_1^2 - 4C_0 C_2}}{2C_2} \tag{5.24}$$

By substituting the root x_c in moment expression of Equation 5.21, the relationship for P-M interaction is obtained as given below:

$$M_u = D_0 + D_1 x_c(P_u) + D_2 x_c^2(P_u) + \frac{D_3[x_c(P_u) - d]}{D - x_c(P_u) - d} \tag{5.25}$$

where the constants, $D_{i=0,1,2,3}$ are given as below:

$$D_0 = \frac{b(d-D)\left[\varepsilon_{c0}\sigma_{c0}(D(\varepsilon_{c0} + 2\varepsilon_{su}) - d\varepsilon_{c0}) + 6(2d - D)p_t\varepsilon_{su}^2\sigma_{s0}\right]}{12\varepsilon_{su}^2}$$

$$D_1 = \frac{b\sigma_{c0}}{6\varepsilon_{su}^2}\left[D\left(\varepsilon_{c0}^2 + 3\varepsilon_{c0}\varepsilon_{su} + 3\varepsilon_{su}^2\right) - d\varepsilon_{c0}(\varepsilon_{c0} + 2\varepsilon_{su})\right]$$

$$D_2 = \frac{-b\sigma_{c0}\left(\varepsilon_{c0}^2 + 4\varepsilon_{c0}\varepsilon_{su} + 6\varepsilon_{su}^2\right)}{12\varepsilon_{su}^2}$$

$$D_3 = \frac{bp_cE_s\varepsilon_{su}}{2}(2d^2 - 3dD + D^2)$$

$$\tag{5.26}$$

In the subdomain $2_b^{(2)}$, steel in compression zone starts yielding while the depth of *plastic kernel* of concrete assumes the same value as given by Equation 5.19. The position of the neutral axis lies in the range $[x_{c,lim}^{(b)}, x_{c,lim}'']$, where $x_{c,lim}''$ is the limit position of neutral axis between the subdomains $2_b^{(2)}$ and 3 for strain in compression steel reaching elastic limit (refer to Figure 5.2, subdomain $2_b^{(2)}$) and is given by

$$x_{c,lim}'' = \left(\frac{\varepsilon_{cu}}{\varepsilon_{su} + \varepsilon_{cu}}\right)(D - d) = \left(\frac{0.0035}{0.01 + 0.0035}\right)(D - d)$$

$$= 0.259(D - d) \qquad \text{for } \varepsilon_{c,max} = \varepsilon_{cu}; \quad \varepsilon_{st} = \varepsilon_{su}$$

$$\tag{5.27}$$

Ultimate axial force and bending moment in subdomain $2_b^{(2)}$ are given by

$$\left\{ \begin{array}{l} P_u = b\left[\displaystyle\int_q^{x_c} \sigma_c(\varepsilon_c(y))\,dy + q\sigma_{c0} + \sigma_{s0}(D - d)(p_c - p_t)\right] \\[20pt] M_u = b\left[\displaystyle\int_q^{x_c} \sigma_c(\varepsilon_c(y))\left(\frac{D}{2} - y\right)dy + \frac{q\sigma_{c0}}{2}(D - q) + \sigma_{s0}(D - d)(p_c + p_t)\left(\frac{D}{2} - d\right)\right] \end{array} \right.$$

$$x_{c,lim}^{(b)} \leq x_c \leq x_{c,lim}''$$

$$\tag{5.28}$$

By solving the expression of axial force in the above equation, depth of neutral axis is determined as

$$x_c = \frac{3P_u\varepsilon_{su} + b(D-d)[\varepsilon_{c0}\sigma_{c0} + 3\varepsilon_{su}\sigma_{s0}(p_t - p_c)]}{b\sigma_{c0}(\varepsilon_{c0} + 3\varepsilon_{su})} \tag{5.29}$$

By substituting in the moment expression of Equation 5.28, P-M relationship for this domain is obtained as

$$M_u = E_0 + E_1 x_c(P_u) + E_2 x_c^2(P_u) \tag{5.30}$$

where the constants $E_{i=0,1,2}$ are given by:

$$E_0 = \frac{b}{12}\left[\frac{(D-d)[\varepsilon_{c0}d - D(\varepsilon_{c0} + 2\varepsilon_{su})]\sigma_{c0}}{\varepsilon_{su}^2} + 6\sigma_{s0}(p_c + p_t)(2d^2 + D^2 - 3dD) \right]$$

$$E_1 = \frac{b\sigma_{c0}}{6\varepsilon_{su}^2}\left[D\left(\varepsilon_{c0}^2 + 3\varepsilon_{c0}\varepsilon_{su} + 3\varepsilon_{su}^2\right) - d\varepsilon_{c0}(\varepsilon_{c0} + 2\varepsilon_{su}) \right] \tag{5.31}$$

$$E_2 = -\frac{b\sigma_{c0}}{12\varepsilon_{su}^2}\left[\varepsilon_{c0}^2 + 4\varepsilon_{c0}\varepsilon_{su} + 6\varepsilon_{su}^2 \right]$$

5.3.2 SUBDOMAINS 3 TO 6b: COLLAPSE CAUSED BY CRUSHING OF CONCRETE

The strain profile for concrete and steel for different subdomains is shown in Figure 5.3. By imposing the respective strain limits in concrete and steel, the position of the neutral axis between the respective limit values can be determined as explained above. However, it is necessary to know that the plastic flow rule shall stand verified in the tensile failure zone initiated by yielding of steel; therefore a detailed mathematical derivation is presented in the above section. For continuity of understanding the limit domains in compression failure zone, a summary of expressions for all ten domains is given in Table 5.1. A detailed procedure can be seen from the literature (see, for example, Chandrasekaran et al. 2008a). The presented summary of expressions may be readily used by designers to identify the damage to cross-section based on strain profile of constitutive materials. Adding to the designer's point of interest, influence of tension and compression reinforcements on the developed P-M interaction domain can also be seen from Chandrasekaran et al. (2008a).

5.4 PLASTIC STRAIN INCREMENT IN DIFFERENT SUBDOMAINS

For a stress state belonging to the yield boundary of P-M curve and moving on the curve, plastic strain increment, in vector form, can be expressed as

$$d\varepsilon_p = \begin{bmatrix} d\varepsilon_{CG} \\ d\phi \end{bmatrix} \tag{5.32}$$

TABLE 5.1

Summary of Expressions for P-M Yield Interaction for Different Subdomains

Sub-domain	x_c	$q(x_c)$	$\varepsilon_{c,max}(x_c)$	$\varepsilon_{st}(x_c)$	$\varepsilon_{sc}(x_c)$	$\lvert\sigma_{st}(x_c)\rvert$	$\lvert\sigma_{sc}(x_c)\rvert$	$P_u(x_c)$	$M_u(x_c)$
1	$\left]-\infty, x_{c,lim}^{(0)}\right]$	0	$\varepsilon_{su}\dfrac{x_c}{D-x_c-d}$	ε_{su}	$\varepsilon_{su}\left(\dfrac{x_c-d}{D-x_c-d}\right)$	σ_{s0}	σ_{s0}	$P_{u,st}$	$M_{u,st}$
$2_a^{(1)}$	$\left[x_{c,lim}^{(0)}, x_{c,lim}^{(a)}\right]$	0	$\varepsilon_{su}\dfrac{x_c}{D-x_c-d}$	ε_{su}	$\varepsilon_{su}\dfrac{x_c-d}{D-x_c-d}$	σ_{s0}	$E_s\lvert\varepsilon_{sc}\rvert$	$P_{u,st}$	$M_{u,st}$
$2_a^{(2)}$	$\left[x_{c,lim}^{(a)}, x'_{c,lim}\right]$	0	$\varepsilon_{su}\dfrac{x_c}{D-x_c-d}$	ε_{su}	$\varepsilon_{su}\dfrac{x_c-d}{D-x_c-d}$	σ_{s0}	$E_s\lvert\varepsilon_{sc}\rvert$	$P_{u,st} + P_{u1,con}(q=0)$	$M_{u,st} + M_{u1,con}(q=0)$
$2_b^{(1)}$	$\left[x'_{c,lim}, x''_{c,lim}\right]$	$x_c - \dfrac{\varepsilon_{c0}}{\varepsilon_{su}}(D-x_c-d)$	$\varepsilon_{su}\dfrac{x_c}{D-x_c-d}$	ε_{su}	$\varepsilon_{su}\dfrac{x_c-d}{D-x_c-d}$	σ_{s0}	$E_s\lvert\varepsilon_{sc}\rvert$	$P_{u,st} + P_{u1,con}$	$M_{u,st} + M_{u1,con}$
$2_b^{(2)}$	$\left[x_{c,lim}^{(b)}, x''_{c,lim}\right]$	$x_c - \dfrac{\varepsilon_{c0}}{\varepsilon_{su}}(D-x_c-d)$	$\varepsilon_{su}\dfrac{x_c}{D-x_c-d}$	ε_{su}	$\varepsilon_{su}\dfrac{x_c-d}{D-x_c-d}$	σ_{s0}	σ_{s0}	$P_{u,st} + P_{u1,con}$	$M_{u,st} + M_{u1,con}$
3	$\left[x''_{c,lim}, x'''_{c,lim}\right]$	$\dfrac{\varepsilon_{cu}-\varepsilon_{c0}}{\varepsilon_{cu}}x_c$	ε_{cu}	$\varepsilon_{cu}\dfrac{(D-x_c-d)}{x_c}$	$\varepsilon_{cu}\dfrac{(x_c-d)}{x_c}$	σ_{s0}	σ_{s0}	$P_{u,st} + P_{u1,con}$	$M_{u,st} + M_{u1,con}$
4	$\left[x'''_{c,lim}, (D-d)\right]$	$\dfrac{\varepsilon_{cu}-\varepsilon_{c0}}{\varepsilon_{cu}}x_c$	ε_{cu}	$\varepsilon_{cu}\dfrac{(D-x_c-d)}{x_c}$	$\varepsilon_{cu}\dfrac{(x_c-d)}{x_c}$	$E_s\lvert\varepsilon_{st}\rvert$	σ_{s0}	$P_{u,st} + P_{u1,con}$	$M_{u,st} + M_{u1,con}$

(Continued)

TABLE 5.1 (CONTINUED)
Summary of Expressions for P-M Yield Interaction for Different Subdomains

Sub-domain	x_c	$q(x_c)$	$\varepsilon_{c,max}(x_c)$	$\varepsilon_{st}(x_c)$	$\varepsilon_{sc}(x_c)$	$|\sigma_{st}(x_c)|$	$|\sigma_{sc}(x_c)|$	$P_u(x_c)$	$M_u(x_c)$
5	$[(D-d), D]$	$\dfrac{\varepsilon_{cu} - \varepsilon_{c0}}{\varepsilon_{cu}} x_c$	ε_{cu}	$\dfrac{\varepsilon_{cu}(D - x_c - d)}{x_c}$	$\dfrac{\varepsilon_{cu}(x_c - d)}{x_c}$	$E_s|\varepsilon_{st}|$	σ_{s0}	$P_{u,st} + P_{u1,con}$	$M_{u,st} + M_{u1,con}$
6a	$[D, x_{c,lim}^{IV}]$	$\dfrac{\varepsilon_{cu} - \varepsilon_{c0}}{\varepsilon_{cu}} D$	$\dfrac{\varepsilon_{cu}\varepsilon_{c0} x_c}{\varepsilon_{cu} x_c - D(\varepsilon_{cu} - \varepsilon_{c0})}$	$\dfrac{\varepsilon_{cu}\varepsilon_{c0}(D - x_c - d)}{\varepsilon_{cu} x_c - D(\varepsilon_{cu} - \varepsilon_{c0})}$	$\dfrac{\varepsilon_{cu}\varepsilon_{c0}(x_c - d)}{\varepsilon_{cu} x_c - D(\varepsilon_{cu} - \varepsilon_{c0})}$	$E_s|\varepsilon_{st}|$	σ_{s0}	$P_{u,st} + P_{u1,con} + P_{u2,con}$	$M_{u,st} + M_{u1,con} + M_{u2,con}$
6b	$[x_{c,lim}^{IV}, +\infty[$	$\dfrac{\varepsilon_{cu} - \varepsilon_{c0}}{\varepsilon_{cu}} D$	$\dfrac{\varepsilon_{cu}\varepsilon_{c0} x_c}{\varepsilon_{cu} x_c - D(\varepsilon_{cu} - \varepsilon_{c0})}$	$\dfrac{\varepsilon_{cu}\varepsilon_{c0}(D - x_c - d)}{\varepsilon_{cu} x_c - D(\varepsilon_{cu} - \varepsilon_{c0})}$	$\dfrac{\varepsilon_{cu}\varepsilon_{c0}(x_c - d)}{\varepsilon_{cu} x_c - D(\varepsilon_{cu} - \varepsilon_{c0})}$	σ_{s0}	σ_{s0}	$P_{u,st} + P_{u1,con} + P_{u2,con}$	$M_{u,st} + M_{u1,con} + M_{u2,con}$

where $\varepsilon_{c,max}(x_c) = \varepsilon_c(x_c, y = 0)$ $P_{u,st} = A_{sc}\sigma_{sc} - A_{st}\sigma_{st} = b(D - d)(p_c\sigma_{sc} - p_t\sigma_{st})$

$$M_{u,st} = (A_{sc}\sigma_{sc} - A_{st}\sigma_{st})\left(\frac{D}{2} - d\right) = b(D - d)(p_c\sigma_{sc} - p_t\sigma_{st})\left(\frac{D}{2} - d\right)$$

$$P_{u1,con} = bq\sigma_{c0} + \frac{b\sigma_{c0}\varepsilon_{c,max}(q - x_c)^2}{3x_c^2\varepsilon_{c0}^2}[3\varepsilon_{c0}x_c + \varepsilon_{c,max}(q - x_c)]$$

$$M_{u1,con} = \frac{bq\sigma_{c0}}{2}(D - q) - \frac{b\sigma_{c0}\varepsilon_{c,max}(q - x_c)^2}{12x_c^2\varepsilon_{c0}^2} \times [2\varepsilon_{c0}x_c(4q + 2x_c - 3D) + \varepsilon_{c,max}(q - x_c)(x_c + 3q - 2D)]$$

$$P_{u2,con} = -\frac{b\sigma_{c0}\varepsilon_{c,max}[3\varepsilon_{c0}x_c + \varepsilon_{c,max}(D - x_c)](D - x_c)^2}{3x_c^2\varepsilon_{c0}^2}$$

$$M_{u2,con} = \frac{b\sigma_{c0}\varepsilon_{c,max}(D - x_c)^2\left[2\varepsilon_{c0}x_c(D - 2x_c) + \varepsilon_{c,max}\left(D^2 - x_c^2\right)\right]}{12x_c^2\varepsilon_{c0}^2}$$

where $d\varepsilon_{CG}$ is strain increment along the axis of the beam evaluated at the CG of the cross-section and $d\phi$ is the curvature increment. It is usual in the theory of plasticity that plastic strain vector increments given by Equation 5.32 shall be represented in the same plane reporting P-M yield domain by placing the axes of $d\varepsilon_{CG}$ and $d\phi$ upon P and M, respectively. The vector $d\varepsilon_p$ is shown connected to the relevant stress point belonging to the P-M yield boundary of every subdomain, as seen in Figure 5.4. Axial strain at CG and curvature in subdomains 1 to $2_b^{(2)}$ are given by

$$\varepsilon_{CG} = \left(\frac{\varepsilon_{su}}{D - xc - d}\right)\left(x_c - \frac{D}{2}\right)$$

(5.33)

$$\phi = \frac{\varepsilon_{su}}{D - xc - d}$$

(5.34)

By solving Equation 5.33 with respect to x_c, we obtain the following relationship:

$$x_c = \frac{D\varepsilon_{su} - 2(D - d)\varepsilon_{CG}}{2(\varepsilon_{su} - \varepsilon_{CG})}$$

(5.35)

Substituting in Equation 5.34, the relationship between curvature and strain at CG is obtained as given below:

$$\phi = \frac{2(\varepsilon_{su} - \varepsilon_{CG})}{D - 2d}$$

(5.36)

The derivative of Equation 5.36 with respect to axial strain increment at CG is given by

$$\frac{d\phi}{d\varepsilon_{CG}} = -\frac{2}{D - 2d}, \quad \alpha_p = \frac{180}{\pi}\left(\pi - \arctan\left|\frac{d\phi}{d\varepsilon_{CG}}\right|\right) \text{ (in deg)}$$

(5.37)

where α_p is the angle between the plastic strain vector and strain axis that assumes a constant value given by Equation 5.37 in subdomains 1 to $2_b^{(2)}$. Strain increment at CG and curvature in subdomains 3 to 5 are given by

$$\varepsilon_{CG} = \varepsilon_{cu}\left(1 - \frac{D}{2x_c}\right)$$

(5.38)

$$\phi = \frac{\varepsilon_{cu}}{xc}$$

(5.39)

By solving Equation 5.38 with respect to x_c and substituting in Equation 5.39, we obtain the relationship between curvature and strain increment at CG as given below:

$$\phi = \frac{2(\varepsilon_{cu} - \varepsilon_{CG})}{D}$$

(5.40)

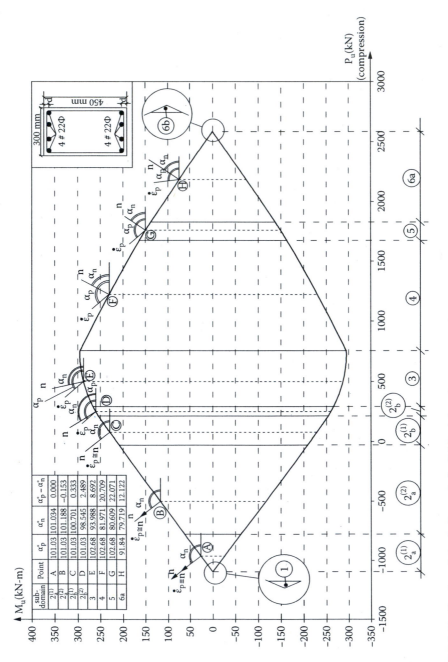

FIGURE 5.4 Verification of plastic flow rule for P-M subdomains.

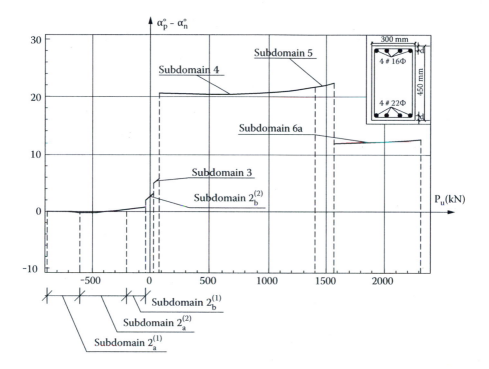

FIGURE 5.5 Verification of plastic flow rule for P-M subdomains (RC beam 300×450 mm with p_c not equal to p_t).

Further, derivative of curvature given by the above equation with respect to axial strain increment at CG is given by

$$\frac{d\phi}{d\varepsilon_{CG}} = -\frac{2}{D}, \qquad \alpha_p = \frac{180}{\pi}\left(\pi - \arctan\left|\frac{d\phi}{d\varepsilon_{CG}}\right|\right) \text{ (in deg)} \qquad (5.41)$$

where, α_p assumes a constant values given by Equation 5.41 in subdomains 3 to 5. In subdomains 6a to 6b, depth of *plastic kernel* of concrete is limited to $(3/7D)$ to limit maximum strain in concrete to its ultimate value. Now, strain at CG and curvature in subdomains 6a to 6b are given by

$$\varepsilon_{CG} = \left[\frac{\varepsilon_{cu}\varepsilon_{c0}}{\varepsilon_{cu}xc - D(\varepsilon_{cu} - \varepsilon_{c0})}\right]\left(x_c - \frac{D}{2}\right) \qquad (5.42)$$

$$\phi = \frac{\varepsilon_{cu}\varepsilon_{c0}}{\varepsilon_{cu}xc - D(\varepsilon_{cu} - \varepsilon_{c0})} \qquad (5.43)$$

By solving Equation 5.42 with respect to x_c and substituting in Equation 5.43, we obtain the relationship between curvature and strain at CG as given below:

$$\phi = \frac{2\varepsilon_{cu}(\varepsilon_{c0} - \varepsilon_{CG})}{D(2\varepsilon_{c0} - \varepsilon_{cu})} \tag{5.44}$$

The derivative of curvature in the above equation with respect to strain at CG is given by

$$\frac{d\phi}{d\varepsilon_{CG}} = \frac{2\varepsilon_{cu}}{D(\varepsilon_{cu} - 2\varepsilon_{c0})}, \qquad \alpha_p = \frac{180}{\pi}\left(\pi - \arctan\left|\frac{d\phi}{d\varepsilon_{CG}}\right|\right) \text{ (in deg)} \tag{5.45}$$

where α_p assumes a constant value given by Equation 5.46 in subdomains 6a to 6b. By summarizing the results, we can write:

$$\frac{d\phi}{d\varepsilon_{CG}} = \begin{cases} \dfrac{2}{2d - D} & \text{subdomains (1) to } (2_b^{(2)}) \\[2mm] -\dfrac{2}{D} & \text{subdomains (3) to (5)} \\[2mm] \dfrac{2\varepsilon_{cu}}{D(\varepsilon_{cu} - 2\varepsilon_{c0})} & \text{subdomains (6a) and (6b)} \end{cases} \tag{5.46}$$

5.5 VERIFICATION OF FLOW RULE

The developed P-M interaction relationships are now verified in different subdomains, both in tension and compression failure zones. In subdomain 1, it may be noted that the ultimate axial force and bending moment are independent of the position of neutral axis, as seen from Equations 5.4 and 5.5. Therefore, verification of flow rule does not apply to this subdomain. In subdomain $2_a^{(1)}$, ultimate moment is given by Equation 5.8, and its derivate with respect to axial force is given by

$$\frac{dM_u}{dP_u} = \frac{D - 2d}{2}, \qquad \alpha_n = \frac{180}{\pi}\left(\frac{\pi}{2} + \arctan\left(\frac{dM_u}{dP_u}\right)\right) \text{ (in deg)} \tag{5.47}$$

where α_n is the angle between the normal to P-M boundary and strain axis, $d\varepsilon_{CG}$. The product of Equations 5.37 and 5.47 gives the following relationship:

$$\left(\frac{dM_u}{dP_u}\right)\cdot\left(\frac{d\phi}{d\varepsilon_{CG}}\right) = -1 \quad \Rightarrow \quad dM d\phi + dP d\varepsilon_{CG} = 0 \tag{5.48}$$

This perfectly verifies the plastic flow rule in subdomain $2_a^{(1)}$. Similarly, verification of the plastic flow rule in other subdomains is carried out and the results are plotted as shown in Figure 5.4. The verification is also illustrated through an example. An RC beam of size 300×450 mm, reinforced with 4#22Φ on the tension and compression side with R_{ck} as 25 N/mm² and f_y as 415 N/mm² is now considered. The P-M boundary showing all the subdomains is plotted in Figure 5.4. Different points, A to H, one on each subdomain from $2_a^{(1)}$ to 6a are identified; angles between the normal to the P-M boundary (α_n) and plastic strain vector (α_p) with respect to $d\varepsilon_{CG}$ axis are computed. It can be seen from Figure 5.4 that the normality rule is well satisfied in subdomains $2_a^{(1)}$ to $2_b^{(1)}$, whereas it is not completely satisfied in subdomains $2_b^{(2)}$ to 6a; it means that the developed P-M interaction relationships are well agreed with the plastic flow rule in the subdomains causing tension failure, with an exception in subdomain $2_b^{(2)}$, since this is the limit boundary between tension and compression failure zones.

In the case of subdomains of compression failure, since the damage is initiated by strain in concrete reaching its limit value leading to crushing of concrete, the flow rule verification fails. In subdomains 6a–6b, strain in concrete is reaching its ultimate limit and the section is becoming more plasticized (see Figure 5.3). Strain profile is rotated about the point Q since ultimate limit strain in concrete is fixed (as imposed by Eurocode); hence, the plastic flow rule cannot be verified since there is no continuity in the strain increment. Table 5.2 shows the numerical values of the angles between the strain axis and normal and tangent of the plastic strain vectors for different subdomains. It can be seen that plastic normality rule qualifies well in subdomains $2_a^{(1)}$–$2_b^{(2)}$, but it does not satisfy completely in subdomains 3 to 6a.

5.6 CONCLUSIONS

A detailed methodology of examining the plastic flow rule in the proposed P-M yield interaction subdomains is presented in this chapter. The mathematically developed P-M interaction model is capable of identifying the damage mechanism of different subdomains in RC sections; damage identification is made on the basis of strain profile of concrete and reinforcing steel. The verified plastic flow is in close agreement with normality in all subdomains of tension failure, while it does not qualify in a few of the subdomains of crushing failure. Also, verification of the plastic flow rule on the proposed P-M interaction relationships is influenced neither by cross-section area of the members nor by variation of tension and compression reinforcements. The developed P-M interaction boundary that is subsequently verified for complete agreement in tension zone, in particular (where failure is initiated by yielding of steel), will enhance the confidence level of structural designers to use the proposed expressions. With the help of the proposed summary of expressions presented in a closed form, it is believed that structural design of new RC buildings and assessment of existing buildings can be performed with better understanding and improved accuracy.

TABLE 5.2
Numerical Values for Different Examples in Subdomains

Sub-domains	Point	P_u(kN)	M_u(kN-m)	x_c (m)	ϕ (rad/m)	ε_{CG}	$\dfrac{d\phi}{d\varepsilon_{CG}}$	$\dfrac{dM_u}{dP_u}$	$\dfrac{d\phi}{d\varepsilon_{CG}}\dfrac{dM_u}{dP_u}$	α_p°	α_n°	$\alpha_p^\circ - \alpha_n^\circ$
1	—	−1096.87	0.00	−2.771	0.0031	−0.0094	−5.128	0.195	−1.000	101.03	101.03	0.00
$2_a^{(1)}$	A	−961.67	26.36	−0.028	0.0223	−0.0056	−5.128	0.195	−1.000	101.03	101.03	0.00
$2_a^{(2)}$	B	−499.59	117.17	0.032	0.0257	−0.0050	−5.128	0.198	−1.014	101.03	101.19	−0.15
$2_b^{(1)}$	C	78.11	229.15	0.078	0.0292	−0.0043	−5.128	0.189	−0.969	101.03	100.70	0.33
$2_b^{(2)}$	D	249.41	260.22	0.097	0.0310	−0.0040	−5.128	0.150	−0.771	101.03	98.55	2.49
3	E	499.69	287.52	0.187	0.0188	−0.0007	−4.444	0.070	−0.310	102.68	93.99	8.69
4	F	1222.13	230.56	0.344	0.0102	0.0012	−4.444	−0.141	0.627	102.68	81.97	20.71
5	G	1759.22	149.46	0.437	0.0080	0.0017	−4.444	−0.165	0.735	102.68	80.61	22.07
6a	H	2184.26	73.98	1.196	0.0020	0.0019	−31.111	−0.181	5.643	91.84	79.72	12.12
6b	—	2578.69	1.02	1.912	0.0012	0.0020	−31.111	−0.161	5.000	91.84	80.87	10.97

$$\frac{d\phi}{d\varepsilon_{CG}} = \begin{cases} \dfrac{2}{2d - D} & \text{subdomains 1 to } 2_b^{(2)} \\[2ex] -\dfrac{2}{D} & \text{subdomains 3 to 5} \\[2ex] \dfrac{2\varepsilon_{cu}}{D(\varepsilon_{cu} - 2\varepsilon_{c0})} & \text{subdomains 6a and 6b} \end{cases}$$

$$\alpha_p^\circ = \frac{180}{\pi}\left[\pi - \arctan\left|\frac{d\phi}{d\varepsilon_{CG}}\right|\right]$$

$$\alpha_n^\circ = \frac{180}{\pi}\left[\frac{\pi}{2} + \arctan\left(\frac{dM_u}{dP_u}\right)\right]$$

Other detailed studies (Chandrasekaran et al. 2008a) conducted by the authors on RC beams to examine the influence of the developed P-M interaction sub-domains are also quite useful for the design engineers and are summarized here for their benefit. However, these studies are not the subject of this chapter and hence are not presented in detail. For increased percentage of tension reinforce-ment (with a fixed percentage of compression reinforcement), the P-M boundary gets elongated along its leading diagonal without influencing the boundary limit of subdomains, causing crushing failure; for a fixed percentage of tension rein-forcement, increase in the percentage of compression reinforcement elongates the P-M boundary along its shorter diagonal without influencing the boundary limit of subdomains, causing tension failure. Increase in the areas of cross-section of the beam show enlargement in the P-M boundary. Influence of material characteristics is also examined by the authors in detail. The results show that increase in yield strength of steel reinforcement enlarges the P-M boundaries nominally, but this nominal enlargement is symmetrical about the axial load axis. Change in charac-teristic compressive strength of concrete in the beams influences P-M boundaries by enlarging the subdomains of crushing failure while those of tension failure are not influenced at all.

APPENDIX: SUMMARY OF P-M RELATIONSHIPS FOR DIFFERENT SUBDOMAINS

The stress-strain limits for the different subdomains are seen in Table 1. The follow-ing summary of expressions is useful to determine the P-M relationships in different subdomains.

Subdomain 1

$$M_u = P_u b \left(\frac{D}{2} - d \right)$$

A(5.1)

Subdomain $2_a^{(1)}$

$$M_u = [P_u + 2b(D-d)p_t\sigma_{s0}] \left(\frac{D}{2} - d \right)$$

A(5.2)

Subdomain $2_a^{(2)}$

$$M_u = B_0 + \left(\frac{1}{D - x_{c3}(P_u) - d} \right) \left[B_1(x_{c3}(P_u)) - d + \frac{B_2 x_{c3}^2(P_u) + B_3 x_{c3}^3(P_u) + B_4 x_{c3}^4(P_u)}{(D - x_{c3}(P_u) - d)} \right]$$

A(5.3)

where the constants, $B_{i=0,1,2,3,4}$ are given by

$$B_0 = \frac{b p_t \sigma_{s0}}{2} (2d^2 - 3dD + D^2)$$

$$B_1 = \frac{1}{2} b(2d^2 - 3dD + D^2) E_s p_c \varepsilon_{su}$$

$$B_2 = \frac{bD(D-d)\varepsilon_{su}\sigma_{c0}}{2\varepsilon_{c0}} \qquad\qquad A(5.4)$$

$$B_3 = \frac{b\varepsilon_{su}\sigma_{c0}}{6\varepsilon_{c0}^2} [2d\varepsilon_{c0} - D(5\varepsilon_{c0} + \varepsilon_{su})]$$

$$B_4 = \frac{b\varepsilon_{su}}{12\varepsilon_{c0}^2} (4\varepsilon_{c0} + \varepsilon_{su})\sigma_{c0}$$

$$x_{c3}(P_u) = \frac{1}{12A_3}\left[-4A_2 - \frac{(2.5198 - 4.3645i)(A_2^2 - 3A_1A_3)}{\lambda} - (1.5874 + 2.7495i)\lambda \right]$$

$$A(5.5)$$

$$\lambda = \left[\sqrt{-4(A_2^2 - 3A_1A_3)^3 + (2A_2^3 - 9A_1A_2A_3 + 27A_3^2A_0)^2} - 2A_2^3 + 9A_1A_2A_3 - 27A_3^2A_0 \right]^{1/3}$$

$$A(5.6)$$

where the constants, $A_{i=0,1,2,3}$, are given by

$$A_0 = -(d-D)^2[P_u + bd E_s p_c \varepsilon_{su} + b(D-d)p_t\sigma_{s0}]$$

$$A_1 = (D-d)[2P_u + bD E_s p_c \varepsilon_{su} + 2b(D-d)p_t\sigma_{s0}]$$

$$A_2 = \frac{-P_u\varepsilon_{c0} + b(d-D)[\varepsilon_{su}(E_s p_c \varepsilon_{s0} - \sigma_{c0}) + p_t\varepsilon_{c0}\sigma_{s0}]}{\varepsilon_{c0}} \qquad A(5.7)$$

$$A_3 = \frac{-b\varepsilon_{su}(3\varepsilon_{c0} + \varepsilon_{su})\sigma_{c0}}{3\varepsilon_{c0}^2}$$

Subdomain $2_b^{(1)}$

$$M_u = D_0 + D_1 x_c(P_u) + D_2 x_c^2(P_u) + \frac{D_3[x_c(P_u) - d]}{D - x_c(P_u) - d} \qquad A(5.8)$$

where $D_{i=0,1,2}$ are constants given by the following relationship:

$$D_0 = \frac{b(d-D)\left[\varepsilon_{c0}\sigma_{c0}(D(\varepsilon_{c0}+2\varepsilon_{su})-d\varepsilon_{c0})+6(2d-D)p_t\varepsilon_{su}^2\sigma_{s0}\right]}{12\varepsilon_{su}^2}$$

$$D_1 = \frac{b\sigma_{c0}}{6\varepsilon_{su}^2}\left[D\left(\varepsilon_{c0}^2+3\varepsilon_{c0}\varepsilon_{su}+3\varepsilon_{su}^2\right)-d\varepsilon_{c0}(\varepsilon_{c0}+2\varepsilon_{su})\right]$$

$$D_2 = \frac{-b\sigma_{c0}\left(\varepsilon_{c0}^2+4\varepsilon_{c0}\varepsilon_{su}+6\varepsilon_{su}^2\right)}{12\varepsilon_{su}^2}$$

A(5.9)

$$D_3 = \frac{bp_cE_s\varepsilon_{su}}{2}(2d^2-3dD+D^2)$$

$$x_c = -\frac{C_1+\sqrt{C_1^2-4C_0C_2}}{2C_2}$$

A(5.10)

where $C_{i=0,1,2}$ are constants given by the following relationship:

$$C_0 = \frac{(D-d)\left[3P_u\varepsilon_{su}+b\left(3dE_sp_c\varepsilon_{su}^2+(D-d)(3p_t\varepsilon_{su}\sigma_{s0}+\varepsilon_{c0}\sigma_{c0})\right)\right]}{3\varepsilon_{su}}$$

$$C_1 = -\frac{3P_u\varepsilon_{su}+b(d-D)[2\varepsilon_{c0}\sigma_{c0}+3\varepsilon_{su}(E_sp_c\varepsilon_{su}+\sigma_{c0}+p_t\sigma_{s0})]}{3\varepsilon_{su}}$$

A(5.11)

$$C_2 = \frac{b\sigma_{c0}}{3}\left(3+\frac{\varepsilon_{c0}}{\varepsilon_{su}}\right)$$

Subdomain $2_b^{(2)}$

$$M_u = E_0 + E_1 x_c(P_u) + E_2 x_c^2(P_u)$$

A(5.12)

where $E_{i=0,1,2}$ are constants given by the following relationships:

$$E_0 = \frac{b}{12}\left[\frac{(D-d)[\varepsilon_{c0}d-D(\varepsilon_{c0}+2\varepsilon_{su})]\sigma_{c0}}{\varepsilon_{su}^2}+6\sigma_{s0}(p_c+p_t)(2d^2+D^2-3dD)\right]$$

$$E_1 = \frac{b\sigma_{c0}}{6\varepsilon_{su}^2}\left[D\left(\varepsilon_{c0}^2+3\varepsilon_{c0}\varepsilon_{su}+3\varepsilon_{su}^2\right)-d\varepsilon_{c0}(\varepsilon_{c0}+2\varepsilon_{su})\right]$$

A(5.13)

$$E_2 = -\frac{b\sigma_{c0}}{12\varepsilon_{su}^2}\left[\varepsilon_{c0}^2+4\varepsilon_{c0}\varepsilon_{su}+6\varepsilon_{su}^2\right]$$

$$x_c = \frac{3P_u \varepsilon_{su} + b(D-d)[\varepsilon_{c0}\sigma_{c0} + 3\varepsilon_{su}\sigma_{s0}(p_t - p_c)]}{b\sigma_{c0}(\varepsilon_{c0} + 3\varepsilon_{su})} \qquad A(5.14)$$

Subdomain 3

$$M_u = \frac{b\sigma_{s0}(2d^2 - 3dD + D^2)(p_t + p_c)}{2}$$

$$+ \frac{[P_u + b\sigma_{s0}(D-d)(p_t - p_c)]}{4b\sigma_{c0}\varepsilon_{cu}(\varepsilon_{c0} - 3\varepsilon_{cu})^2}$$

$$\times \Big[2bD\sigma_{c0}\varepsilon_{cu}(\varepsilon_{c0} - 3\varepsilon_{cu})^2 - 3\varepsilon_{cu}\left(\varepsilon_{c0}^2 - 4\varepsilon_{c0}\varepsilon_{cu} + 6\varepsilon_{cu}^2\right)$$

$$[P_u + b\sigma_{s0}(D-d)(p_t - p_c)] \Big] \qquad A(5.15)$$

Subdomains 4 and 5

$$M_u = G_0 + G_1 x_c(P_u) + G_2 x_c^2(P_u) + \frac{G_3}{x_c(P_u)} \qquad A(5.16)$$

where the constants $G_{i=0,1,2,3}$ are given by

$$G_0 = -\frac{b}{2}(2d^2 - 3dD + D^2)(E_s p_t \varepsilon_{cu} - p_c \sigma_{s0})$$

$$G_1 = \frac{bD\sigma_{c0}(3\varepsilon_{cu} - \varepsilon_{c0})}{6\varepsilon_{cu}}$$

$$G_2 = \frac{-b\sigma_{c0}\left(\varepsilon_{c0}^2 - 4\varepsilon_{c0}\varepsilon_{cu} + 6\varepsilon_{cu}^2\right)}{12\varepsilon_{cu}^2} \qquad A(5.17)$$

$$G_3 = \frac{b}{2}(d-D)^2(D-2d)E_s p_t \varepsilon_{cu}$$

$$x_c = \frac{-F_1 + \sqrt{F_1^2 - 4F_0 F_2}}{2F_2} \qquad A(5.18)$$

$$F_0 = -b(d-D)^2 E_s p_t \varepsilon_{su}$$

$$F_1 = b(D-d)[E_s p_t \varepsilon_{cu} + p_c \sigma_{s0}] - P_u \qquad A(5.19)$$

$$F_2 = \sigma_{c0} b\left(1 - \frac{\varepsilon_{c0}}{3\varepsilon_{cu}}\right)$$

Subdomain 6a

$$M_u = J_0 + \frac{J_1 + J_1 x_c(P_u)}{D(\varepsilon_{c0} - \varepsilon_{cu}) + x_c(P_u)\varepsilon_{cu}} + \frac{J_3 + J_4 x_c(P_u) + J_5 x_c^2(P_u)}{[D(\varepsilon_{c0} - \varepsilon_{cu}) + x_c(P_u)\varepsilon_{cu}]^2} \qquad A(5.20)$$

where the constants $J_{i=0,1,2,3,4,5}$ are given by

$$J_0 = \frac{b}{2}\left[\frac{D^2\varepsilon_{c0}(\varepsilon_{cu} - \varepsilon_{c0})\sigma_{c0}}{\varepsilon_{cu}^2} + (2d^2 - 3dD + D^2)p_c\sigma_{s0}\right]$$

$$J_1 = \frac{b}{2}E_s p_t \varepsilon_{c0}\varepsilon_{cu}(D - 2d)(d - D)^2$$

$$J_2 = \frac{b}{2}(2d^2 - 3dD + D^2)E_s p_t \varepsilon_{c0}\varepsilon_{cu}$$

$$J_3 = \frac{bD^4\varepsilon_{c0}\sigma_{c0}\left[\varepsilon_{c0}^2(5\varepsilon_{c0} - 16\varepsilon_{cu}) + 6\varepsilon_{cu}^2(3\varepsilon_{c0} - \varepsilon_{cu})\right]}{12\varepsilon_{cu}^2} \qquad A(5.21)$$

$$J_4 = \frac{bD^3\varepsilon_{c0}\sigma_{c0}(\varepsilon_{c0} - \varepsilon_{cu})^2}{\varepsilon_{cu}}$$

$$J_5 = \frac{bD^2\varepsilon_{c0}(\varepsilon_{c0} - \varepsilon_{cu})\sigma_{c0}}{2}$$

$$x_c = \frac{-H_1 + \sqrt{H_1^2 - 4H_0 H_2}}{2H_2} \qquad A(5.22)$$

$$H_0 = D\left[3(\varepsilon_{cu} - \varepsilon_{c0})\varepsilon_{cu}\left(b\varepsilon_{cu}^2\varepsilon_{c0}E_s p_t(d - D)^2 + D(\varepsilon_{c0} - \varepsilon_{cu})(P_u + b(\sigma_{s0}p_c d\right.\right.$$
$$\left.\left. - D(\sigma_{c0} + p_c\sigma_{s0}))\right)\right) + bD^2\varepsilon_{c0}^3\sigma_{c0}\right]$$

$$H_1 = 3\varepsilon_{cu}^2[b(D - d)E_s p_t \varepsilon_{c0}(D(\varepsilon_{c0} - 2\varepsilon_{cu}) + \varepsilon_{cu}d)$$
$$+ 2D(\varepsilon_{cu} - \varepsilon_{c0})(P_u + b(p_c\sigma_{s0}d - D(p_c\sigma_{so} + \sigma_{c0})))]$$

$$H_2 = -3\varepsilon_{cu}^3[P_u + (D - d)(E_s p_t \varepsilon_{c0} + p_c\sigma_{s0})]$$

$$A(5.23)$$

Subdomain 6b

$$M_u = \frac{D(\varepsilon_{c0} - 2\varepsilon_{cu})}{4\varepsilon_{cu}}(P_u - bD\sigma_{c0}) - \frac{b(D - d)\sigma_{s0}}{4\varepsilon_{cu}} \qquad A(5.24)$$
$$\times [4d\varepsilon_{cu}(p_c - p_t) + D(\varepsilon_{c0}(p_c + p_t) - 4p_c\varepsilon_{cu})]$$

6 Computer Coding for Collapse Multipliers

6.1 INTRODUCTION

Reinforced concrete building frames with different geometry are analyzed, and bounds for collapse loads are determined. The computer coding used for this will be discussed in this section. With reference to numerical studies discussed in Chapter 4, the cases considered for the analysis are (1) single bay–single story regular frame, (2) single bay–double story regular frame, (3) single bay–single story with unequal column length, (4) four bay–two story regular frame, (5) six bay–three story irregular frame, (6) six bay–three story regular frame, and (7) five bay–ten story regular frame. Figures 6.1 to 6.6 show the elevation of the building frames considered for the analysis. All building frames are comprised of (1) 450 mm square RC columns, reinforced with 12Φ25 and lateral ties of 8 mm at 200 c/c; (2) 300 × 450 mm RC beam, reinforced with 4Φ22 as tensile and compression steel with shear stirrups of 10 mm at 250 c/c; as well as (3) 125-mm-thick RC slab. M25 mix and high-yield-strength deformed bars (Fe 415) are used. All building frames consisting of 4 m bay widths and 4 m story heights are assumed to be located in Zone V (IS 1893, 2002) with soil condition as "medium" type. Seismic weight at each floor is computed using IS code (IS 1893, 2002), and the base shear is distributed along the height of the building. Live load of equivalent magnitude is considered to act at the midspan of the beam, and lateral loads, computed from the base shear, are assumed to act at each floor level. With the proposed expressions for P-M interaction and moment-rotation, beams and columns are modeled.

6.2 COMPUTER CODING FOR COLLAPSE MULTIPLIERS

The computer coding listed below is compatible with Wolfram Mathematica (Version 6.0.0) for Windows platform.

Quit[ ]

Units for Length (m), Moment (kN-m), Force (kN)

p0 = Live load (as per admissible loading clause): 2.5 kN/sq.m

Story height H, length of beam L, breadth of beam b, overall depth of beam D0, density of concrete γ, size of square column D1, slab thickness s

6.2.1 SINGLE BAY–SINGLE STORY REGULAR FRAME

H=4; L=4; p0=2.5; b=0.3; D0=0.45; γ=25; D1=0.45; s=0.125;

Live load

q1=(1/2*L*L*p0)/2

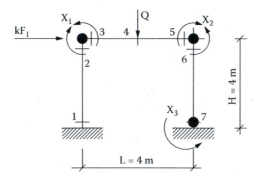

FIGURE 6.1 Single bay–single story regular frame.

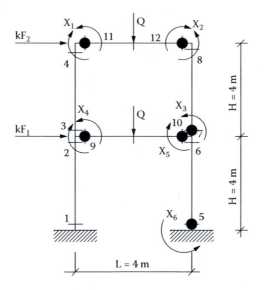

FIGURE 6.2 Single bay–two story regular frame.

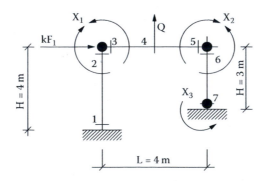

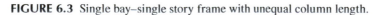

FIGURE 6.3 Single bay–single story frame with unequal column length.

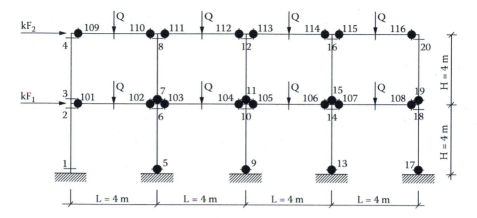

FIGURE 6.4 Four bay–two story regular frame.

10

Dead load for slab

q2=(1/2*L*L*s* γ)/2

12.5

Dead load for finishes

q3=(1/2*L*L*1)/2

4

Dead load for beam

q4=(b*D0*L*γ)/L

3.375

Total dead load on the beam and slab, B1

B1=(1/2)*L*q1*0.5+(1/2)*L*q2+(1/2)*q3*L+q4*L

56.5

Total dead load on the column, Fc

Fc=2*D1*D1*H*γ

40.5

Seismic mass, M

M=(2*B1+Fc)/9.81

15.6473

Spectral ordinate (as per IS 1893)

Sa=((1+15*T)*(UnitStep[T]-UnitStep[T-0.1])+2.50*(UnitStep[T-0.1]-UnitStep[T-0.55])+1.36/T*(UnitStep[T-0.55]-UnitStep[T-4.00]))

(1.36(-UnitStep[-4+T]+UnitStep[-0.55+T]))/T+2.5(-UnitStep[-0.55+T]+UnitStep[-0.1+T])+(1+15 T) (-UnitStep[-0.1+T]+UnitStep[T])

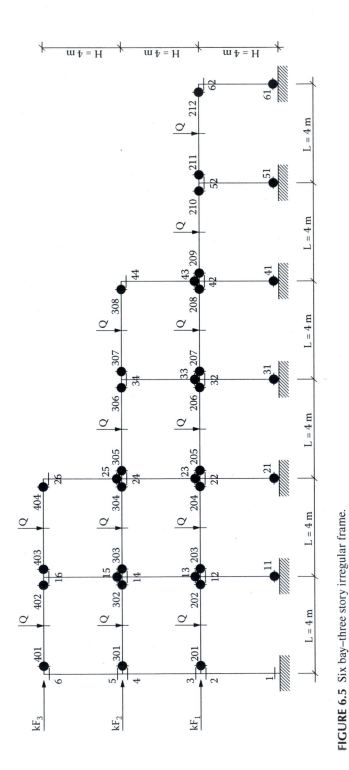

FIGURE 6.5 Six bay–three story irregular frame.

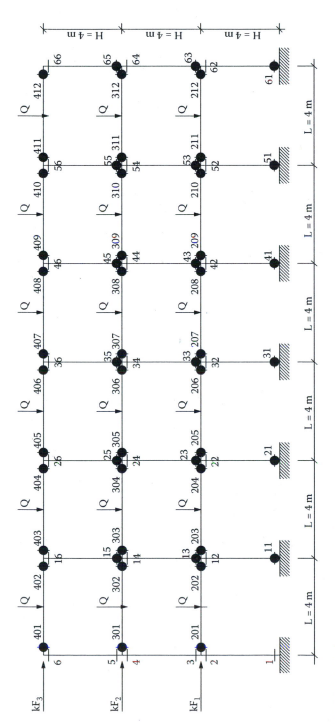

FIGURE 6.6 Six bay–three story regular frame.

Graphic representation of the spectrum

Plot[Sa,{T,0,4},PlotRange →{0,3}]

(The plot thus obtained in the computer screen can be seen in Image 6.1.)

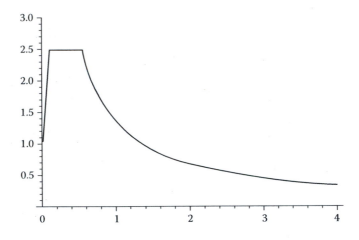

IMAGE 6.1 Response spectrum.

Base shear constants

Z0=0.36; I0=1; R0=5; T=0.075*(H)^0.75

0.212132

Acceleration coefficient for calculating base shear

Ah=Z0/2*I0/R0*Sa

0.09

Base shear

Vb=Ah*M*9.81

13.815

Ultimate bending moment for beam and column

Mb=214.45; Mc=265.06;

Kinematic multiplier, kc [Equation 4.4 of *Seismic Design Aids*]

kc=Simplify[(2*Mb+2*Mc)/(Vb*H)]

17.3547

Static multiplier, ks [Equation 4.8 of *Seismic Design Aids*]

ks=Simplify[Mb/(Vb/2*H/2)]

15.523

6.2.2 SINGLE BAY–TWO STORY REGULAR FRAME

```
Quit[ ]
H=4; L=4; p0=2.5; b=0.3; D0=0.45; γ=25; D1=0.45; s=0.125;
Live load
q1=(1/2*L*L*p0)/2
10
Dead load for slab
q2=(1/2*L*L*s* γ )/2
12.5
Dead load for finishes
q3=(1/2*L*L*1)/2
4
Dead load for beam
q4=(b*D0*L* γ )/L
3.375
Total dead load on the beam and slab, B1
B1=(1/2)*L*q1*0.5+(1/2)*L*q2+(1/2)*q3*L+q4*L
56.5
Total dead load on the column, Fc
Fc=2*D1*D1*H*γ
40.5
Seismic mass for first floor, M1
M1=(2*B1+Fc)/9.81
15.6473
Seismic mass for second floor, M2
M2=(2*B1+Fc)/9.81
15.6473
Total seismic mass
M=M1+M2
31.2946
Spectral ordinate (as per IS 1893)
Sa=((1+15*T)*(UnitStep[T]-UnitStep[T-0.1])+2.50*(UnitStep[T-0.1]-
UnitStep[T-0.55])+1.36/T*(UnitStep[T-0.55]-UnitStep[T-4.00]))
```

(1.36 (-UnitStep[-4+T]+UnitStep[-0.55+T]))/T+2.5 (-UnitStep[-0.55+T]+UnitStep[-0.1+T])+(1+15 T) (-UnitStep[-0.1+T]+UnitStep[T])

Graphic representation of the spectrum

Plot[Sa,{T,0,4},PlotRange →{0,3}]

Base shear constants

Z0=0.36; I0=1; R0=5; T=0.075*(2*H)^0.75

0.356762

Acceleration coefficient for calculating base shear

Ah=Z0/2*I0/R0*Sa

0.09

Base shear

Vb=Ah*M*9.81

27.63

Seismic forces at each floor (starting from the ground floor)

F1=Vb*(M1*(H^2))/(M1*(H^2)+M2*((2*H)^2))

F2=Vb*(M2*((2*H)^2))/(M1*(H^2)+M2*((2*H)^2))

5.526

22.104

Ultimate bending moment for beam and column

Mb=214.45; Mc=265.06;

Kinematic multiplier, kc [Equation 4.4 of *Seismic Design Aids*]

kc=Simplify[(4*Mb+2*Mc)/(F1*H+F2*2*H)]

6.97672

Static multiplier, ks [Equation 4.8 of Design Aids]

ks=Simplify[(2*Mb)/((F1+F2)/2*H+F1/2*H)]

6.46791

6.2.3   SINGLE BAY–SINGLE STORY FRAME WITH UNEQUAL COLUMN LENGTH

Quit[ ]

H=4; L=4; H1=3; p0=2.5; b=0.3; D0=0.45; γ =25; D1=0.45; s=0.125;

Live load

ql=(1/2*L*L*p0)/2

10

Dead load for slab

q2=(1/2*L*L*s*γ)/2

12.5

Dead load for finishes

q3=(1/2*L*L*1)/2

4

Dead load for beam

q4=(b*D0*L*γ)/L

3.375

Total dead load on the beam and slab, B1

B1=(1/2)*L*q1*0.5+(1/2)*L*q2+(1/2)*q3*L+q4*L

56.5

Total dead load on the column, Fc

Fc=D1*D1*H*γ+(H1/2+2)*D1*D1*γ

37.9688

Seismic mass

M=(2*B1+Fc)/9.81

15.3893

Spectral ordinate (as per IS 1893)

Sa=((1+15*T)*(UnitStep[T]-UnitStep[T-0.1])+2.50*(UnitStep[T-0.1]-UnitStep[T-0.55])+1.36/T*(UnitStep[T-0.55]-UnitStep[T-4.00]))

(1.36(-UnitStep[-4+T]+UnitStep[-0.55+T]))/T+2.5(-UnitStep[-0.55+T]+UnitStep[-0.1+T])+(1+15 T)(-UnitStep[-0.1+T]+UnitStep[T])

Graphic representation of the spectrum

Plot[Sa,{T,0,4},PlotRange→{0,3}]

Base shear constants

Z0=0.36; I0=1; R0=5; T=0.075*(H)^0.75

0.212132

Acceleration coefficient for calculating base shear

Ah=Z0/2*I0/R0*Sa

0.09

Base shear

Vb=Ah*M*9.81

13.5872

Ultimate bending moment for beam and column

Mb=214.45; Mc=265.06;

Kinematic multiplier, kc [Equation 4.4 of *Seismic Design Aids*]

kc=Simplify[(((1+H/H1)*Mb+(1+H/H1)*Mc)/(Vb*H)]

20.5866

6.2.4   FOUR BAY–TWO STORY REGULAR FRAME

Quit[ ]

H=4; L=4; p0=2.5; b=0.3; D0=0.45; γ =25; D1=0.45; s=0.125;

Live load

q1=(1/2*L*L*p0)/2

10

Dead load for slab

q2=(1/2*L*L*s*γ)/2

12.5

Dead load for finishes

q3=(1/2*L*L*1)/2

4

Dead load for beam

q4=(b*D0*L*γ)/L

3.375

Total dead load on the beam and slab, B1

B1=(1/2)*L*q1*0.5+(1/2)*L*q2+(1/2)*q3*L+q4*L

56.5

Total dead load on the column, Fc

Fc=5*D1*D1*H*γ

101.25

Seismic mass for first floor, M1

M1=(8*B1+Fc)/9.81

56.3965

Seismic mass for second floor, M2

M2=(8*B1+Fc)/9.81

56.3965

Total seismic mass

M=M1+M2

112.793

Spectral ordinate (as per IS 1893)

Sa=((1+15*T)*(UnitStep[T]-UnitStep[T-0.1])+2.50*(UnitStep[T-0.1]-UnitStep[T-0.55])+1.36/T*(UnitStep[T-0.55]-UnitStep[T-4.00]))

(1.36(-UnitStep[-4+T]+UnitStep[-0.55+T]))/T+2.5(-UnitStep[-0.55+T]+UnitStep[-0.1+T])+(1+15 T) (-UnitStep[-0.1+T]+UnitStep[T])

Graphic representation of the spectrum

Plot[Sa,{T,0,4},PlotRange →{0,3}]

Base shear constants

Z0=0.36; I0=1; R0=5; T=0.075*(2*H)^0.75

0.356762

Acceleration coefficient for calculating base shear

Ah=Z0/2*I0/R0*Sa

0.09

Base shear

Vb=Ah*M*9.81

99.585

Seismic forces at each floor (starting from the ground floor)

F1=Vb*(M1*(H^2))/(M1*(H^2)+M2*((2*H)^2))

F2=Vb*(M2*((2*H)^2))/(M1*(H^2)+M2*((2*H)^2))

19.917

79.668

Ultimate bending moment for beam and column

Mb=214.45; Mc=265.06;

kc=Simplify[(16*Mb+5*Mc)/(F1*H+F2*2*H)]

6.63378

ks=Simplify[(5*Mb)/((F1+F2)/2*H+F1/2*H)]

4.48633

6.2.5   Six Bay–Three Story Irregular Frame

Quit[ ]

H=4; L=4; p0=2.5; b=0.3; D0=0.45; γ=25; D1=0.45; s=0.125;

Live load

q1=(1/2*L*L*p0)/2

10

Dead load for slab

$q2=(1/2*L*L*s*\gamma)/2$

12.5

Dead load for finishes

$q3=(1/2*L*L*1)/2$

4

Dead load for beam

$q4=(b*D0*L*\gamma)/L$

3.375

Total dead load on the beam and slab, B1

$B1=(1/2)*L*q1*0.5+(1/2)*L*q2+(1/2)*q3*L+q4*L$

56.5

Total dead load on the column, Fc

$Fc=D1*D1*H*\gamma$

20.25

Seismic mass for first floor, M1

$M1=(12*B1+7*Fc)/9.81$

83.5627

Seismic mass for second floor, M2

$M2=(8*B1+5*Fc)/9.81$

56.3965

Seismic mass for third floor, M3

$M3=(4*B1+3*Fc)/9.81$

29.2304

Total seismic mass

$M=M1+M2+M3$

169.19

Spectral ordinate (as per IS 1893)

$Sa=((1+15*T)*(UnitStep[T]-UnitStep[T-0.1])+2.50*(UnitStep[T-0.1]-UnitStep[T-0.55])+1.36/T*(UnitStep[T-0.55]-UnitStep[T-4.00]))$

$(1.36(-UnitStep[-4+T]+UnitStep[-0.55+T]))/T+2.5(-UnitStep[-0.55+T]+UnitStep[-0.1+T])+(1+15T)(-UnitStep[-0.1+T]+UnitStep[T])$

Graphic representation of the spectrum

$Plot[Sa,\{T,0,4\},PlotRange\rightarrow\{0,3\}]$

Base shear constants

$Z0=0.36; I0=1; R0=5; T=0.075*(3*H)^{0.75}$

0.483556

Acceleration coefficient for calculating base shear

Ah=Z0/2*I0/R0*Sa

0.09

Base shear

Vb=Ah*M*9.81

149.378

Seismic forces at each floor (starting from the ground floor)

F1=Vb*(M1*(H^2))/(M1*(H^2)+M2*((2*H)^2)+M3*((3*H)^2))

F2=Vb*(M2*((2*H)^2))/(M1*(H^2)+M2*((2*H)^2)+M3*((3*H)^2))

F3=Vb*(M3*((3*H)^2))/(M1*(H^2)+M2*((2*H)^2)+M3*((3*H)^2))

21.8139

58.8888

68.6748

Ultimate bending moment for beam and column

Mb=214.45; Mc=265.06;

Kinematic multiplier, kc [Equation 4.4 of *Seismic Design Aids*]

kc=Simplify[(24*Mb+7*Mc)/(F1*H+F2*2*H+F3*3*H)]

5.06503

6.2.6   SIX BAY–THREE STORY REGULAR FRAME

Quit[ ]

H=4; L=4; p0=2.5; b=0.3; D0=0.45; γ=25; D1=0.45; s=0.125;

Live load

q1=(1/2*L*L*p0)/2

10

Dead load for slab

q2=(1/2*L*L*s*γ)/2

12.5

Dead load for finishes

q3=(1/2*L*L*1)/2

4

Dead load for beam

q4=(b*D0*L*γ)/L

3.375

Total dead load on the beam and slab, B1

B1=(1/2)*L*q1*0.5+(1/2)*L*q2+(1/2)*q3*L+q4*L

56.5

Total dead load on the column, Fc

Fc=D1*D1*H*γ

20.25

Seismic mass for first floor, M1

M1=(12*B1+7*Fc)/9.81

83.5627

Seismic mass for second floor, M2

M2=(12*B1+7*Fc)/9.81

83.5627

Seismic mass for third floor, M3

M3=(12*B1+7*Fc)/9.81

83.5627

Total seismic mass

M=M1+M2+M3

250.688

Spectral ordinate (as per IS 1893)

Sa=((1+15*T)*(UnitStep[T]-UnitStep[T-0.1])+2.50*(UnitStep[T-0.1]-UnitStep[T-0.55])+1.36/T*(UnitStep[T-0.55]-UnitStep[T-4.00]))

(1.36(-UnitStep[-4+T]+UnitStep[-0.55+T]))/T+2.5(-UnitStep[-0.55+T]+UnitStep[-0.1+T])+(1+15 T)(-UnitStep[-0.1+T]+UnitStep[T])

Graphic representation of the spectrum

Plot[Sa,{T,0,4},PlotRange→{0,3}]

Base shear constants

Z0=0.36; I0=1; R0=5; T=0.075*(3*H)^0.75

0.483556

Acceleration coefficient for calculating base shear

Ah=Z0/2*I0/R0*Sa

0.09

Base shear

Vb=Ah*M*9.81

221.332

Seismic forces at each floor (starting from the ground floor)

F1=Vb*(M1*(H^2))/(M1*(H^2)+M2*((2*H)^2)+M3*((3*H)^2))

F2=Vb*(M2*((2*H)^2))/(M1*(H^2)+M2*((2*H)^2)+M3*((3*H)^2))

F3=Vb*(M3*((3*H)^2))/(M1*(H^2)+M2*((2*H)^2)+M3*((3*H)^2))

15.8095

63.2379

142.285

Ultimate bending moment for beam and column

Mb=214.45; Mc=265.06

Kinematic multiplier, kc [Equation 4.4 of *Seismic Design Aids*]

kc=Simplify[(36*Mb+7*Mc)/(F1*H+F2*2*H+F3*3*H)]

4.20617

Static multiplier, ks [Equation 4.8 of *Seismic Design Aids*]

ks=Simplify[(7*Mb)/((F1+F2+F3)/2*H+(F1+F2)/2*H)]

2.49875

6.2.7 FIVE BAY–TEN STORY REGULAR FRAME

Quit[ ]

H=4; L=4; p0=2.5; b=0.3; D0=0.45; γ=25; D1=0.45; s=0.125;

Live load

q1=(1/2*L*L*p0)/2

10

Dead load for slab

q2=(1/2*L*L*s*γ)/2

12.5

Dead load for finishes

q3=(1/2*L*L*1)/2

4

Dead load for beam

q4=(b*D0*L*γ)/L

3.375

Total dead load on the beam and slab, B1

B1=(1/2)*L*q1*0.5+(1/2)*L*q2+(1/2)*q3*L+q4*L

56.5

Total dead load on the column, Fc

Fc=D1*D1*4*γ

20.25

Seismic mass for each floor

M1=(10*B1+6*Fc)/9.81;

M2=(10*B1+6*Fc)/9.81;

M3=(10*B1+6*Fc)/9.81;

M4=(10*B1+6*Fc)/9.81;

M5=(10*B1+6*Fc)/9.81;

M6=(10*B1+6*Fc)/9.81;

M7=(10*B1+6*Fc)/9.81;

M8=(10*B1+6*Fc)/9.81;

M9=(10*B1+6*Fc)/9.81;

M10=(10*B1+6*Fc)/9.81;

Total seismic mass

M=M1+M2+M3+M4+M5+M6+M7+M8+M9+M10

699.796

Spectral ordinate (as per IS 1893)

Sa=((1+15*T)*(UnitStep[T]-UnitStep[T-0.1])+2.50*(UnitStep[T-0.1]-UnitStep[T-0.55])+1.36/T*(UnitStep[T-0.55]-UnitStep[T-4.00]))

(1.36 (-UnitStep[-4+T]+UnitStep[-0.55+T]))/T+2.5 (-UnitStep[-0.55+T]+UnitStep[-0.1+T])+(1+15 T) (-UnitStep[-0.1+T]+UnitStep[T])

Graphic representation of the spectrum

Plot[Sa,{T,0,4},PlotRange →{0,3}]

Base shear constants

Z0=0.36; I0=1; R0=5; T=0.075*(10*H)^0.75

1.19291

Acceleration coefficient for calculating base shear

Ah=Z0/2*I0/R0*Sa

0.0410426

Base shear

Vb=Ah*M*9.81

281.758

Seismic forces at each floor (starting from the ground floor)

F1=Vb*(M1*(H^2))/(M1*(H^2)+M2*((2*H)^2)+M3*((3*H)^2)+M4*((4*H)^2)
+M5*((5*H)^2)+M6*((6*H)^2)+M7*((7*H)^2)+M8*((8*H)^2)+M9*((9*H)^2)
+M10*((10*H)^2))

F2=Vb*(M2*((2*H)^2))/(M1*(H^2)+M2*((2*H)^2)+M3*((3*H)^2)+M4*((4*H)^2)+M5*((5*H)^2)+M6*((6*H)^2)+M7*((7*H)^2)+M8*((8*H)^2)+M9*((9*H)^2)+M10*((10*H)^2))

F3=Vb*(M3*((3*H)^2))/(M1*(H^2)+M2*((2*H)^2)+M3*((3*H)^2)+M4*((4*H)^2)+M5*((5*H)^2)+M6*((6*H)^2)+M7*((7*H)^2)+M8*((8*H)^2)+M9*((9*H)^2)+M10*((10*H)^2))

F4=Vb*(M4*((4*H)^2))/(M1*(H^2)+M2*((2*H)^2)+M3*((3*H)^2)+M4*((4*H)^2)+M5*((5*H)^2)+M6*((6*H)^2)+M7*((7*H)^2)+M8*((8*H)^2)+M9*((9*H)^2)+M10*((10*H)^2))

F5=Vb*(M5*((5*H)^2))/(M1*(H^2)+M2*((2*H)^2)+M3*((3*H)^2)+M4*((4*H)^2)+M5*((5*H)^2)+M6*((6*H)^2)+M7*((7*H)^2)+M8*((8*H)^2)+M9*((9*H)^2)+M10*((10*H)^2))

F6=Vb*(M6*((6*H)^2))/(M1*(H^2)+M2*((2*H)^2)+M3*((3*H)^2)+M4*((4*H)^2)+M5*((5*H)^2)+M6*((6*H)^2)+M7*((7*H)^2)+M8*((8*H)^2)+M9*((9*H)^2)+M10*((10*H)^2))

F7=Vb*(M7*((7*H)^2))/(M1*(H^2)+M2*((2*H)^2)+M3*((3*H)^2)+M4*((4*H)^2)+M5*((5*H)^2)+M6*((6*H)^2)+M7*((7*H)^2)+M8*((8*H)^2)+M9*((9*H)^2)+M10*((10*H)^2))

F8=Vb*(M8*((8*H)^2))/(M1*(H^2)+M2*((2*H)^2)+M3*((3*H)^2)+M4*((4*H)^2)+M5*((5*H)^2)+M6*((6*H)^2)+M7*((7*H)^2)+M8*((8*H)^2)+M9*((9*H)^2)+M10*((10*H)^2))

F9=Vb*(M9*((9*H)^2))/(M1*(H^2)+M2*((2*H)^2)+M3*((3*H)^2)+M4*((4*H)^2)+M5*((5*H)^2)+M6*((6*H)^2)+M7*((7*H)^2)+M8*((8*H)^2)+M9*((9*H)^2)+M10*((10*H)^2))

F10=Vb*(M10*((10*H)^2))/(M1*(H^2)+M2*((2*H)^2)+M3*((3*H)^2)+M4*((4*H)^2)+M5*((5*H)^2)+M6*((6*H)^2)+M7*((7*H)^2)+M8*((8*H)^2)+M9*((9*H)^2)+M10*((10*H)^2))

0.731838

2.92735

6.58654

11.7094

18.296

26.3462

35.8601

46.8376

59.2789

73.1838

Ultimate bending moment for beam and column

Mb=214.45; Mc=265.06;

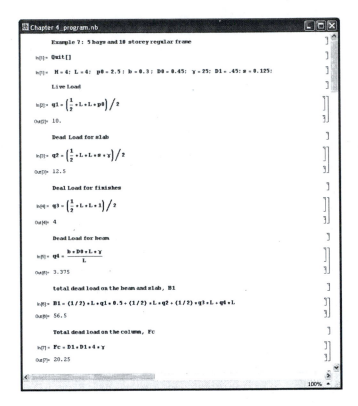

IMAGE 6.2 Five bay–ten story regular frame; coding in Mathematica_page 1.

Kinematic multiplier, kc [Equation 4.4 of *Seismic Design Aids*]

kc=Simplify[(100*Mb+6*Mc)/(F1*H+F2*2*H+F3*3*H+F4*4*H+F5*5*H+F6*6*H+F7*7*H+F8*8*H+F9*9*H+F10*10*H)]

2.60133

Static multiplier, ks [Equation 4.8 of *Seismic Design Aids*]

ks=Simplify[(6*Mb)/((F1+F2+F3+F4+F5+F6+F7+F8+F9+F10)/2*H+(F1+F2+F3+F4+F5+F6+F7+F8+F9)/2*H)]

1.31207

(The computer coding as appears on the screen can be seen in Images 6.2 to 6.6.)

6.2.8 GENERAL PROCEDURE FOR REGULAR FRAMES WITH M BAYS–N STORIES

This section provides the coding for obtaining the collapse multipliers (both static and kinematic) for regular frames with the number of bays and stories of the user's choice.

Quit[ ]

H=4; L=4; p0=2.5; b=0.3; D0=0.45; γ =25; D1=0.45; s=0.125;

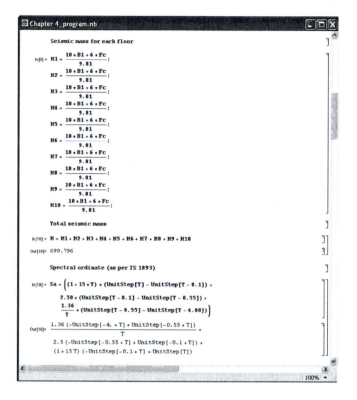

IMAGE 6.3 Five bay–ten story regular frame; coding in Mathematica_page 2.

Live load

q1=(1/2*L*L*p0)/2

10

Dead load for slab

q2=(1/2*L*L*s*γ)/2

12.5

Dead load for finishes

q3=(1/2*L*L*1)/2

4

Dead load for beam

q4=(b*D0*L*γ)/L

3.375

Total dead load on the beam and slab, B1

B1=(1/2)*L*q1*0.5+(1/2)*L*q2+(1/2)*q3*L+q4*L

56.5

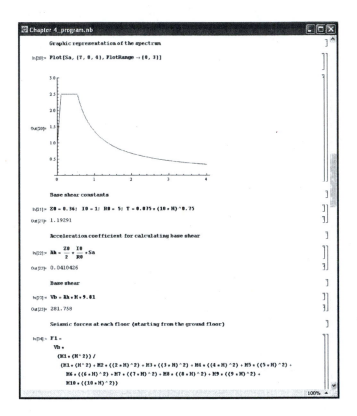

IMAGE 6.4 Five bay–ten story regular frame; coding in Mathematica_page 3.

Total dead load on the column, Fc

Fc=D1*D1*4*γ

20.25

Seismic mass for generic ith floor

Mi=((m+1)*Fc+2*m*B1)/9.81

0.101937 (113. m+20.25 (1+m))

Total seismic mass

M= Mi*(n)

0.101937 (113. m+20.25 (1+m)) n

Spectral ordinate (as per IS 1893)

Sa=((1+15*T)*(UnitStep[T]-UnitStep[T-0.1])+2.50*(UnitStep[T-0.1]-UnitStep[T-0.55])+1.36/T*(UnitStep[T-0.55]-UnitStep[T-4.00]))

(1.36 (-UnitStep[-4+T]+UnitStep[-0.55+T]))/T+2.5 (-UnitStep[-0.55+T]+UnitStep[-0.1+T])+(1+15 T) (-UnitStep[-0.1+T]+UnitStep[T])

```
Chapter 4_program.nb *

F2 =
Vb * (M2 * ((2 * H) ^2)) /
    (M1 * (H^2) + M2 * ((2 * H)^2) + M3 * ((3 * H)^2) + M4 * ((4 * H)^2) + M5 * ((5 * H)^2) +
    M6 * ((6 * H)^2) + M7 * ((7 * H)^2) + M8 * ((8 * H)^2) + M9 * ((9 * H)^2) + M10 * ((10 * H)^2))
F3 =
Vb * (M3 * ((3 * H)^2)) /
    (M1 * (H^2) + M2 * ((2 * H)^2) + M3 * ((3 * H)^2) + M4 * ((4 * H)^2) + M5 * ((5 * H)^2) +
    M6 * ((6 * H)^2) + M7 * ((7 * H)^2) + M8 * ((8 * H)^2) + M9 * ((9 * H)^2) + M10 * ((10 * H)^2))
F4 =
Vb * (M4 * ((4 * H)^2)) /
    (M1 * (H^2) + M2 * ((2 * H)^2) + M3 * ((3 * H)^2) + M4 * ((4 * H)^2) + M5 * ((5 * H)^2) +
    M6 * ((6 * H)^2) + M7 * ((7 * H)^2) + M8 * ((8 * H)^2) + M9 * ((9 * H)^2) + M10 * ((10 * H)^2))
F5 =
Vb * (M5 * ((5 * H)^2)) /
    (M1 * (H^2) + M2 * ((2 * H)^2) + M3 * ((3 * H)^2) + M4 * ((4 * H)^2) + M5 * ((5 * H)^2) +
    M6 * ((6 * H)^2) + M7 * ((7 * H)^2) + M8 * ((8 * H)^2) + M9 * ((9 * H)^2) + M10 * ((10 * H)^2))
F6 =
Vb * (M6 * ((6 * H)^2)) /
    (M1 * (H^2) + M2 * ((2 * H)^2) + M3 * ((3 * H)^2) + M4 * ((4 * H)^2) + M5 * ((5 * H)^2) +
    M6 * ((6 * H)^2) + M7 * ((7 * H)^2) + M8 * ((8 * H)^2) + M9 * ((9 * H)^2) + M10 * ((10 * H)^2))
F7 =
Vb * (M7 * ((7 * H)^2)) /
    (M1 * (H^2) + M2 * ((2 * H)^2) + M3 * ((3 * H)^2) + M4 * ((4 * H)^2) + M5 * ((5 * H)^2) +
    M6 * ((6 * H)^2) + M7 * ((7 * H)^2) + M8 * ((8 * H)^2) + M9 * ((9 * H)^2) + M10 * ((10 * H)^2))
F8 =
Vb * (M8 * ((8 * H)^2)) /
    (M1 * (H^2) + M2 * ((2 * H)^2) + M3 * ((3 * H)^2) + M4 * ((4 * H)^2) + M5 * ((5 * H)^2) +
    M6 * ((6 * H)^2) + M7 * ((7 * H)^2) + M8 * ((8 * H)^2) + M9 * ((9 * H)^2) + M10 * ((10 * H)^2))
F9 =
Vb * (M9 * ((9 * H)^2)) /
    (M1 * (H^2) + M2 * ((2 * H)^2) + M3 * ((3 * H)^2) + M4 * ((4 * H)^2) + M5 * ((5 * H)^2) +
    M6 * ((6 * H)^2) + M7 * ((7 * H)^2) + M8 * ((8 * H)^2) + M9 * ((9 * H)^2) + M10 * ((10 * H)^2))
F10 =
Vb * (M10 * ((10 * H)^2)) /
    (M1 * (H^2) + M2 * ((2 * H)^2) + M3 * ((3 * H)^2) + M4 * ((4 * H)^2) + M5 * ((5 * H)^2) +
    M6 * ((6 * H)^2) + M7 * ((7 * H)^2) + M8 * ((8 * H)^2) + M9 * ((9 * H)^2) + M10 * ((10 * H)^2))

                                                            100%
```

IMAGE 6.5 Five bay–ten story regular frame; coding in Mathematica_page 4.

Graphic representation of the spectrum

Plot[Sa,{T,0,4},PlotRange_{0,3}]

Base shear constants

Z0=0.36; I0=1; R0=5; T=0.075*(n*H)^0.75

$0.212132\ n^{0.75}$

Acceleration coefficient for calculating base shear

Ah=Z0/2*I0/R0*Sa

$0.036\ ((6.4111\ (-\text{UnitStep}[-4.+0.212132\ n^{0.75}]+\text{UnitStep}[-0.55+0.212132\ n^{0.75}]))/n^{0.75}+2.5\ (-\text{UnitStep}[-0.55+0.212132\ n^{0.75}]+\text{UnitStep}[-0.1+0.212132\ n^{0.75}])+(1+3.18198\ n^{0.75})\ (-\text{UnitStep}[-0.1+0.212132\ n^{0.75}]+\text{UnitStep}[0.212132\ n^{0.75}]))$

Base shear

Vb=Ah*M*9.81

$0.036\ (113.\ m+20.25\ (1+m))\ n\ ((6.4111\ (-\text{UnitStep}[-4.+0.212132\ n^{0.75}]+\text{UnitStep}[-0.55+0.212132\ n^{0.75}]))/n^{0.75}+2.5\ (-\text{UnitStep}[-0.55+0.212132\ n^{0.75}]+\text{UnitStep}[-0.1+0.212132\ n^{0.75}])+(1+3.18198\ n^{0.75})\ (-\text{UnitStep}[-0.1+0.212132\ n^{0.75}]+\text{UnitStep}[0.212132\ n^{0.75}]))$

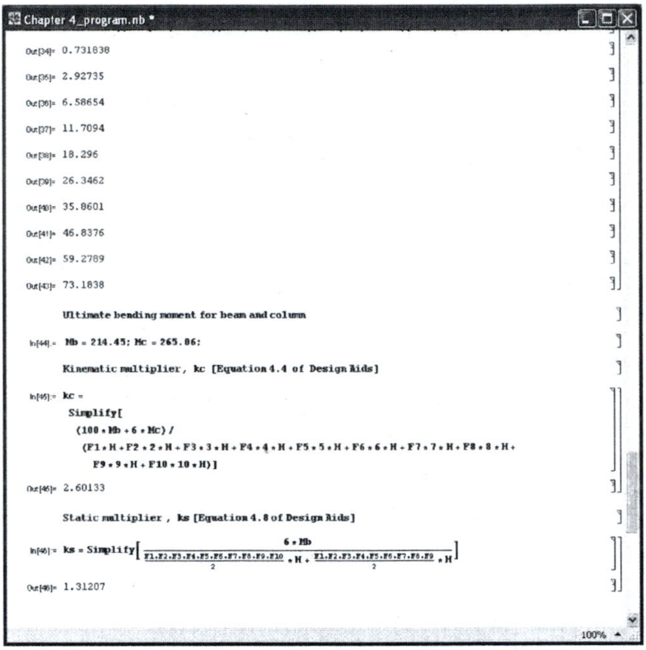

IMAGE 6.6 Five bay–ten story regular frame; coding in Mathematica_page 5.

Seismic forces at each floor (starting from the ground floor)

$$F_i = Vb * \frac{W_i * (i * H)^\wedge 2}{\sum_{i=1}^{n} W_i * (i * H)^\wedge 2}$$

1/((1+n) (1+2 n))0.216 i^2 (113. m+20.25 (1+m)) ((6.4111 (-UnitStep[-4.+0.212132 n$^{0.75}$]+UnitStep[-0.55+0.212132 n$^{0.75}$]))/n$^{0.75}$+2.5 (-UnitStep[-0.55+0.212132 n$^{0.75}$]+UnitStep[-0.1+0.212132 n$^{0.75}$])+(1+3.18198 n$^{0.75}$) (-UnitStep[-0.1+0.212132 n$^{0.75}$]+UnitStep[0.212132 n$^{0.75}$]))

Base shear

$$Vb* = \sum_{i=1}^{n} F_i$$

0.036 (113. m+20.25 (1+m)) n ((6.4111 (-UnitStep[-4.+0.212132 n$^{0.75}$]+UnitStep[-0.55+0.212132 n$^{0.75}$]))/n$^{0.75}$+2.5 (-UnitStep[-0.55+0.212132 n$^{0.75}$]+UnitStep[-0.1+0.212132 n$^{0.75}$])+(1+3.18198 n$^{0.75}$) (-UnitStep[-0.1+0.212132 n$^{0.75}$]+UnitStep[0.212132 n$^{0.75}$]))

Kinematic multiplier, kc [Equation 4.4 of *Seismic Design Aids*]

$$k_c = \text{FullSimplify}\left[\frac{(n*(2*m)*M_b + (m+1)*M_c)}{\sum\limits_{i=1}^{n} F_i * i * H}\right]$$

(4.62963 (1. +2. n) (Mc+m Mc+2. m Mb n))/((20.25 +133.25 m) $n^{1.25}$ (1. +n) (-6.4111 UnitStep[-4.+0.212132 $n^{0.75}$]+(6.4111 -2.5 $n^{0.75}$) UnitStep[-0.55+0.212132 $n^{0.75}$]+(1.5 $n^{0.75}$-3.18198 $n^{1.5}$) UnitStep[-0.1+0.212132 $n^{0.75}$]+($n^{0.75}$+3.18198 $n^{1.5}$) UnitStep[0.212132 $n^{0.75}$]))

Static multiplier, ks [Equation 4.8 of *Seismic Design Aids*]

$$k_s = \text{FullSimplify}\left[\frac{((m+1)*2*M_b)}{\left(\left(\sum\limits_{i=1}^{n} F_i\right) + \left(\sum\limits_{i=1}^{n-1} F_i\right)\right)*H}\right]$$

((0.5 +0.5 m) Mb (0.5 +n) (1. +n))/((20.25 +133.25 m) $n^{0.25}$ (0.036 +0.072 n^2) (-6.4111 UnitStep[-4+0.212132 $n^{0.75}$]+(6.4111 -2.5 $n^{0.75}$) UnitStep[-0.55+0.212132 $n^{0.75}$]+(1.5 $n^{0.75}$-3.18198 $n^{1.5}$) UnitStep[-0.1+0.212132 $n^{0.75}$]+($n^{0.75}$+3.18198 $n^{1.5}$) UnitStep[0.212132 $n^{0.75}$]))

Ultimate bending moment for beam and column

Mb=214.45; Mc=265.06;

Kinematic multiplier for example cases

$k_c/.\{m \to 1, n \to 1\}$

$k_c/.\{m \to 1, n \to 2\}$

$k_c/.\{m \to 4, n \to 2\}$

$k_c/.\{m \to 6, n \to 3\}$

$k_c/.\{m \to 5, n \to 10\}$

17.3547

6.97672

6.63378

4.20617

2.60133

Static multiplier for example cases

$k_c/.\{m \to 1, n \to 1\}$

$k_s/.\{m \to 1, n \to 2\}$

$k_s/.\{m \to 4, n \to 2\}$

$k_s/.\{m \to 6, n \to 3\}$

$k_s/.\{m \to 5, n \to 10\}$

15.523

6.46791

4.48633

2.49875

1.31207

Base shear for example cases

$V_b/.\{m \to 1, n \to 1\}$

$V_b/.\{m \to 1, n \to 2\}$

$V_b/.\{m \to 4, n \to 2\}$

$V_b/.\{m \to 6, n \to 3\}$

$V_b/.\{m \to 5, n \to 10\}$

13.815

27.63

99.585

221.332

281.758

$Plot[\{k_c/.m \to 1, k_s/.m \to 1\}, \{n, 1, 10\}]$

(The plot thus obtained on the computer screen can be seen in Image 6.7.)

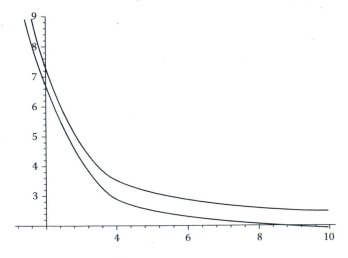

IMAGE 6.7 Collapse multipliers for regular frame (m bays, n stories).

6.2.9 COMPUTER CODING TO COMPUTE STATIC COLLAPSE MULTIPLIERS (LINGO)

Computer coding used to obtain static collapse multipliers by an alternative method (Sforza 2002; Student LINGO 2005) is given for two example cases: (1) single bay–single story regular frame, and (2) single bay–single story irregular frame. The coding can be seen in Images 6.8 and 6.9, respectively.

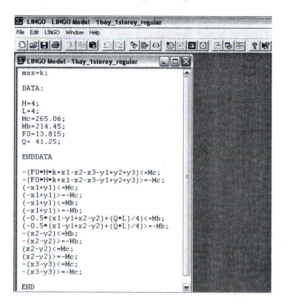

IMAGE 6.8 Single bay–single story regular frame; coding in LINGO.

IMAGE 6.9 Single bay–single story frame with unequal column length; coding in LINGO.

6.3   PROCEDURE TO PERFORM PUSHOVER ANALYSIS

In this section, a detailed procedure to perform nonlinear static pushover analysis is presented. A five bay–ten story regular frame in reinforced concrete is considered as an example case. The building frame consists of structural elements as follows: (1) 450 mm square RC columns, reinforced with 12Φ25 and lateral ties of 8 mm at 200 c/c; (2) 300 × 450 mm RC beam, reinforced with 4Φ22 as tensile and compression steel with shear stirrups of 10 mm at 250 c/c; and (3) 125-mm-thick RC slab. The concrete mix is M25 and the reinforcing steel used is high-yield-strength deformed bars, Fe 415. The building frame consists of 4 m bay width and 4 m story height, with no structural and geometric irregularities and assumed to be located in Zone V (IS 1893, 2002) with soil condition as "medium" type. Using the proposed expressions for P-M interaction and moment-rotation, presented in Chapters 1 and 2, respectively, beams and columns are modeled. Figure 6.7 shows the P-M interaction details for the beam hinges to be used in the model. The P-M interaction domains are traced using the summary of expressions given in Chapter 1. Figure 6.8 shows the moment rotation for the beam hinges, which are plotted using the expressions given in Chapter 3. Similarly, P-M interaction details and moment-rotation for column hinges are shown in Figures 6.9 and 6.10, respectively. The building frame is modeled in SAP2000, version 10.1.2 advanced, using the geometric and structural details as mentioned above. In the following section, a step-by-step approach for performing pushover analysis of the building model is presented.

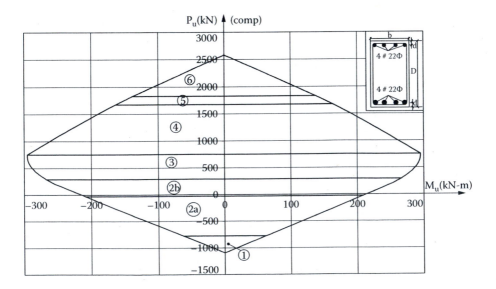

FIGURE 6.7 P-M interaction for beam hinges.

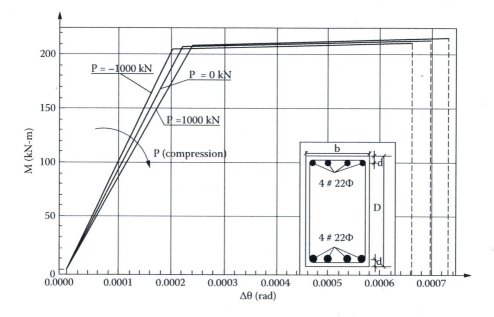

FIGURE 6.8 Moment-rotation for beam hinges.

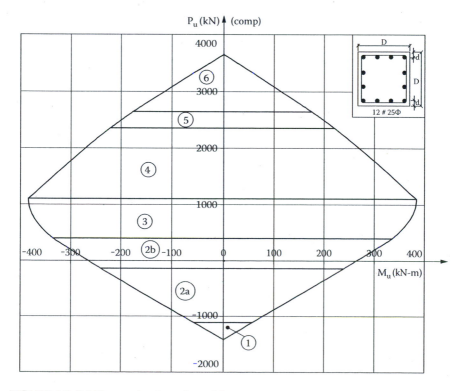

FIGURE 6.9 P-M interaction for column hinges.

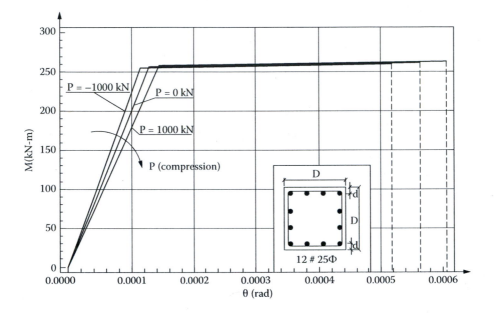

FIGURE 6.10 Moment-rotation for column hinges.

6.3.1 Step-by-Step Approach Using SAP2000

Step 1: Select New Model from the pull-down menu. (Image 6.10)

Step 2: Select the 2D frame with n stories and m bays. Set the units to kN-m. (Image 6.11)

Step 3: Fix the dimensions of the frame. (Image 6.12)

Step 4: The 2D frame is prepared and displayed. (Image 6.13)

IMAGE 6.10 New model from the pull-down menu.

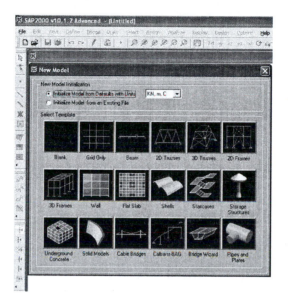

IMAGE 6.11 2D frame with n stories and m bays.

Step 5: Save the file as *ten story bldg_new*. It is necessary to mark the tip node with the desired label to monitor the pushover curve at this node. Select the tip node, and in the menu, select Edit – Change Label; enter *roof top*. (Image 6.14)

Step 6: Display labels of members. Click View on the menu bar; set Display Options – Frames/Cables/Tendons. The default labeling in SAP is as follows: The first bottom column is numbered as 1, and the numbering increases

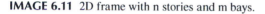

IMAGE 6.12 Fixing the dimensions of the frame.

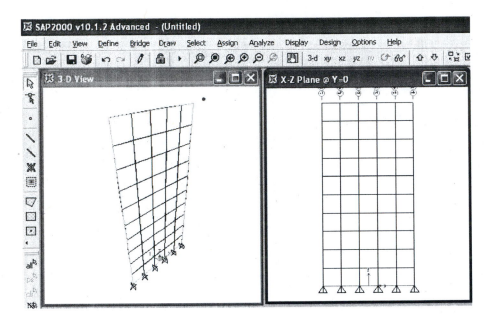

IMAGE 6.13 Display of the prepared 2D frame.

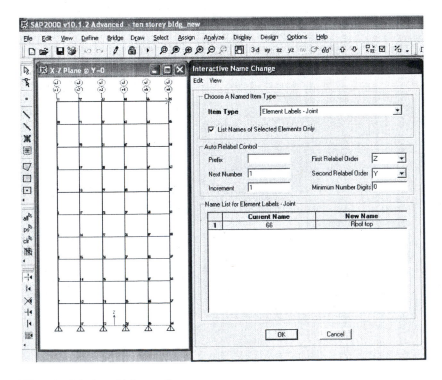

IMAGE 6.14 Display of joint *roof top*.

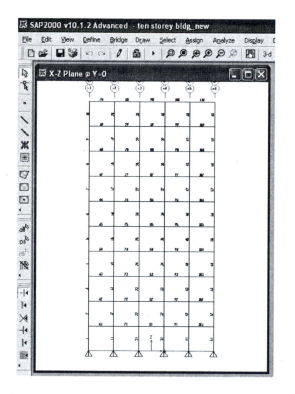

IMAGE 6.15 Numbering of beams.

along the height; this is repeated for other column lines. The numbering of the beams is then assigned (automatically) from the lower floor (left member) and increases along the height. (Image 6.15)

Step 7: Change labels for columns. Go to Edit – Change Labels – Element Labels – Frame. The columns are labeled as IC0J (the prefix I denotes floor number and J denotes column line). One can use Excel program to rename the labels quickly. (Image 6.16)

Step 8: Change labels for beams. Go to Edit – Change Labels – Element Labels – Frame. The beams are labeled as IB0J (the prefix I denotes floor number and J denotes beam line). (Image 6.17)

Step 9: To assign the fixed supports to the columns at the base, select the column joints at the base. Go to Assign – Joint – Restraints. (Image 6.18)

Step 10: Material properties for reinforced concrete. (Image 6.19)

Step 11: Section properties – beams. Go to Menu – Define – Frame Section and enter the details. (Image 6.20)

Step 12: Section properties – columns. Go to Menu – Define – Frame Section and enter the details. (Image 6.21)

Step 13: Assign beams and columns of the frame with appropriate sections. Go to Menu – Assign – Frame/Cable/Tendon – Frame Sections. (Image 6.22)

Step 14: Define nonlinear hinge properties for beam hinges. Go to Menu – Define – Hinge Properties. (Image 6.23)

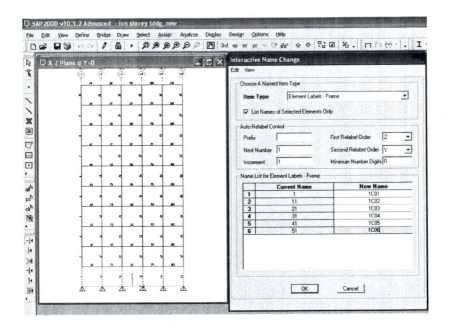

IMAGE 6.16 Numbering of columns.

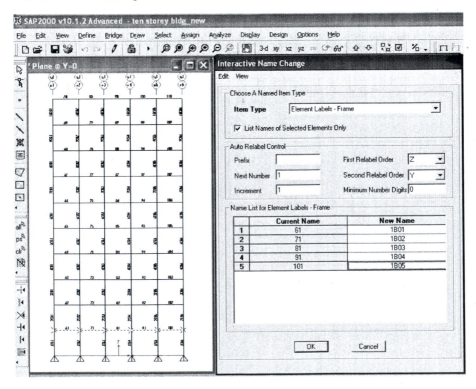

IMAGE 6.17 Changing labels for beams.

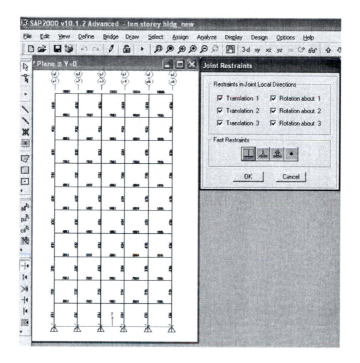

IMAGE 6.18 Joint constraints.

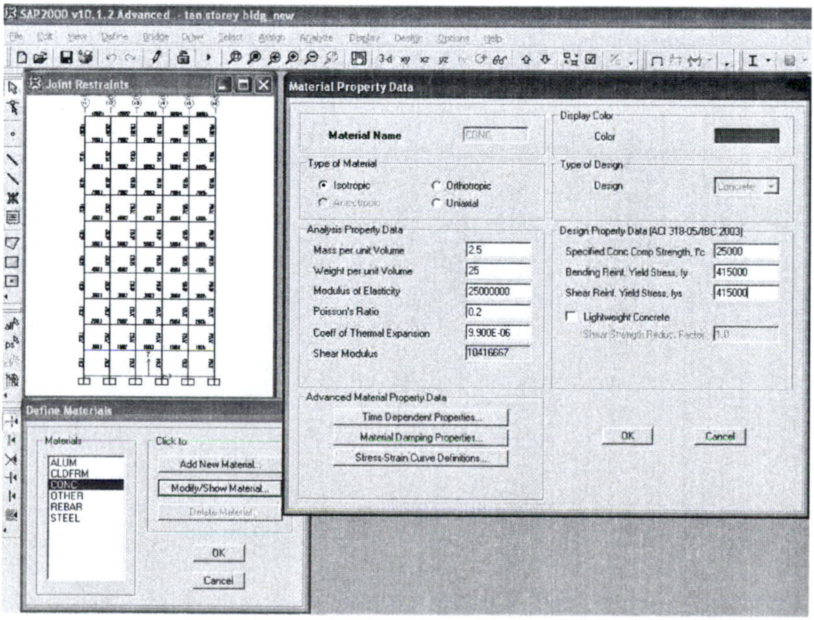

IMAGE 6.19 Material property data.

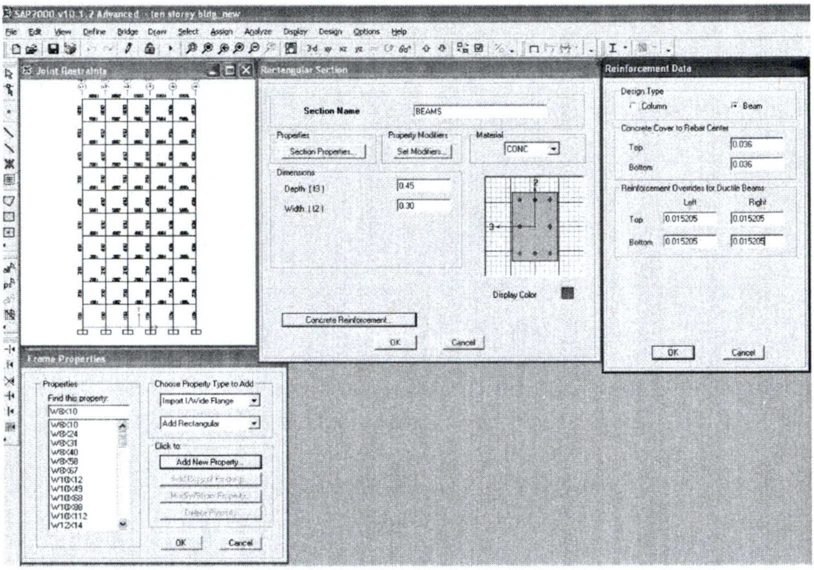

IMAGE 6.20 Section property for beams.

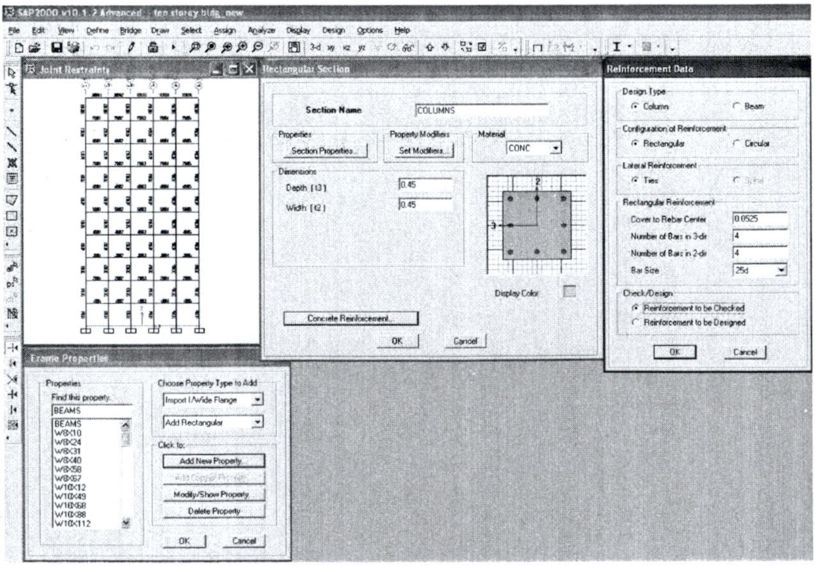

IMAGE 6.21 Section property for columns.

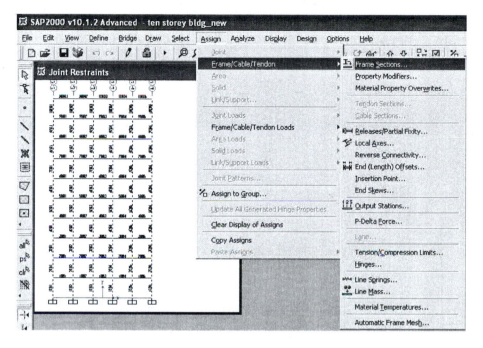

IMAGE 6.22 Assigning frame sections.

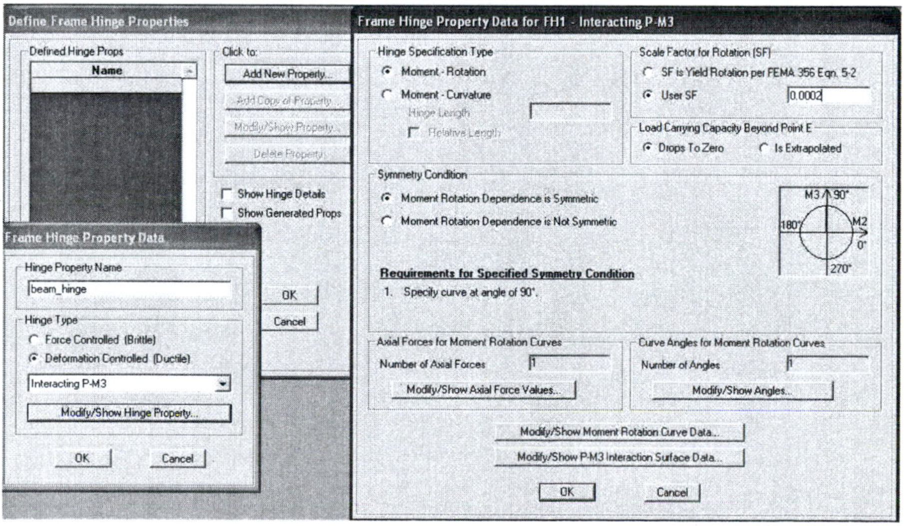

IMAGE 6.23 Nonlinear hinge properties for beams, step 1.

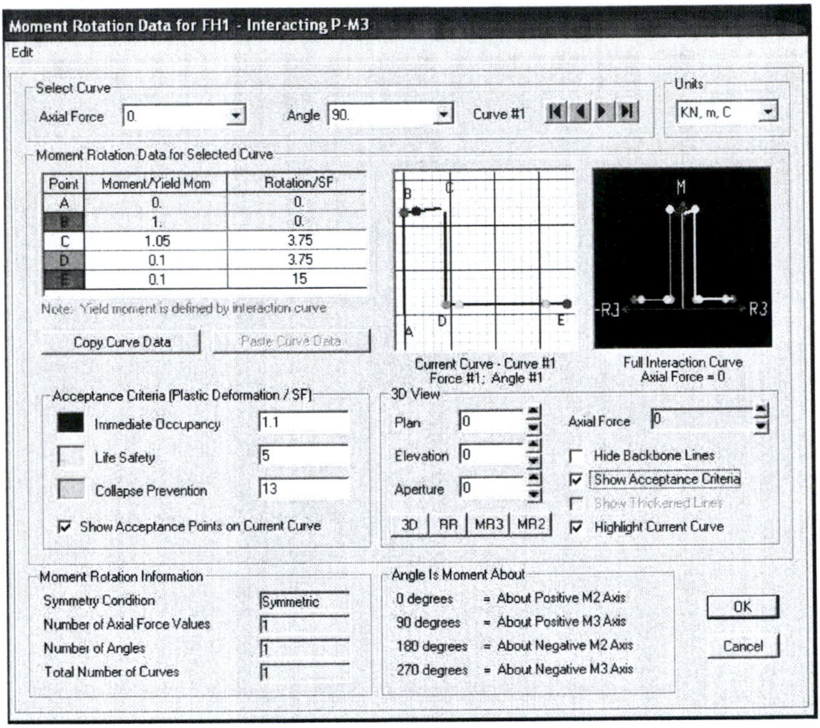

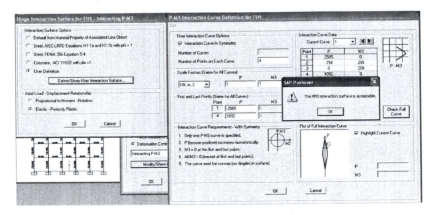

IMAGE 6.24 Nonlinear hinge properties for beams, step 2.

Step 15: Define nonlinear hinge properties for beam hinges, continued. Go to
   Menu – Define – Hinge Properties. (Image 6.24)
Step 16: Define nonlinear hinge properties for beam hinges, continued. Go to
   Menu – Define – Hinge Properties. (Image 6.25)
Step 17: Define nonlinear hinge properties for column hinges. Go to Menu –
   Define – Hinge Properties. (Image 6.26)

IMAGE 6.25 Nonlinear hinge properties for beams, step 3.

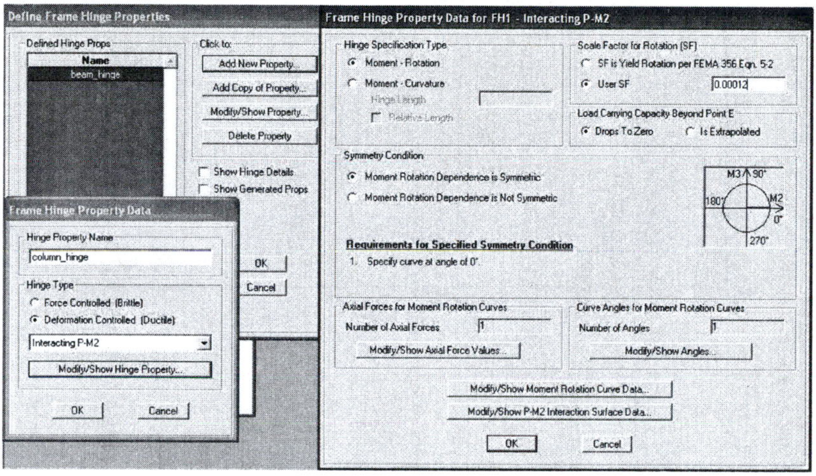

IMAGE 6.26 Nonlinear hinge properties for columns, step 1.

Step 18: Define nonlinear hinge properties for column hinges, continued. Go
to Menu – Define – Hinge Properties. (Image 6.27)

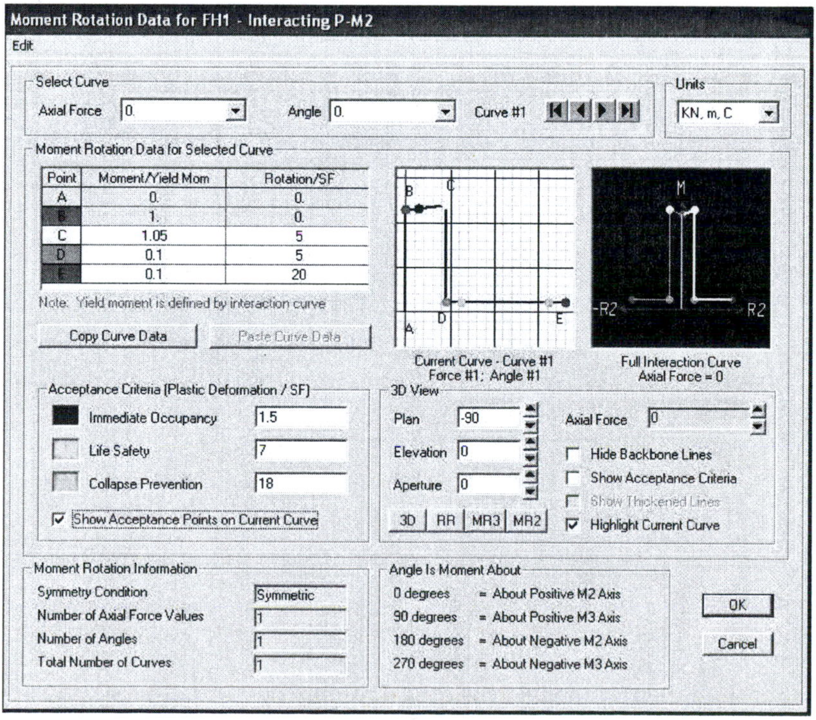

IMAGE 6.27 Nonlinear hinge properties for columns, step 2.

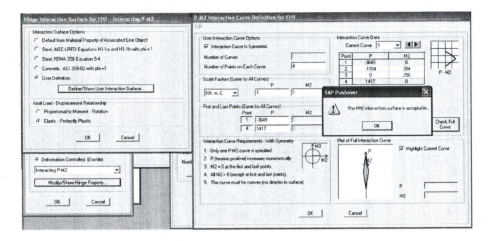

IMAGE 6.28 Nonlinear hinge properties for columns, step 3.

Step 19: Define nonlinear hinge properties for column hinges, continued. Go
to Menu – Define – Hinge Properties. (Image 6.28)

Step 20: Mass source. Go to Menu – Define – Mass Source. (Image 6.29)

Step 21: Assign end length offset to ensure rigidity of connections between
beams and columns. Select all the members and joints, in total. Go to Menu –
Assign – Frame/Cable/Tendon – End (length) Offsets. (Image 6.30)

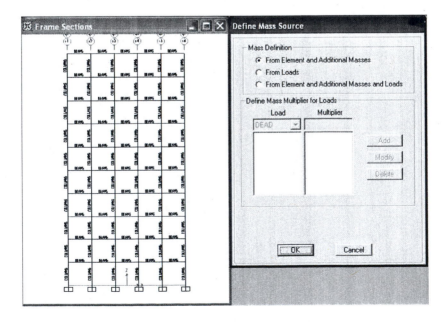

IMAGE 6.29 Defining mass source.

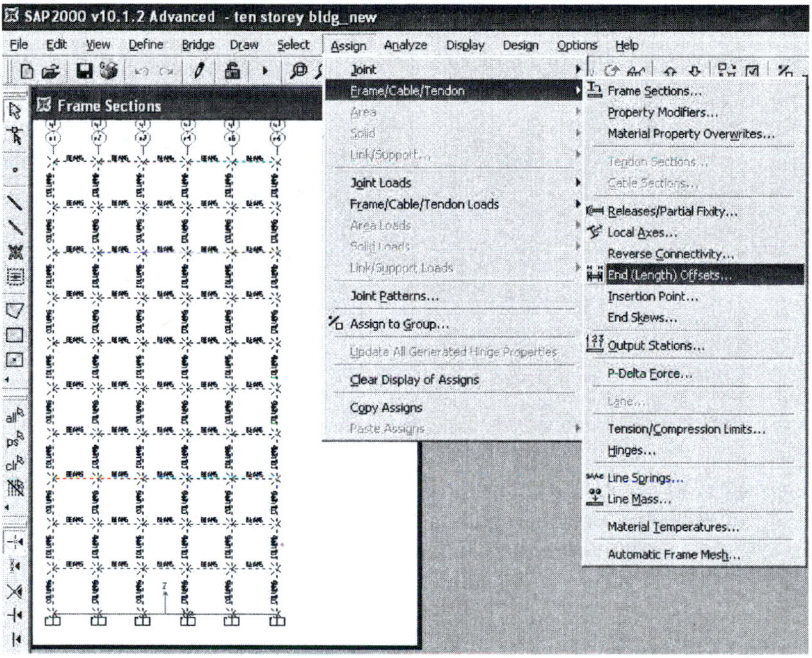

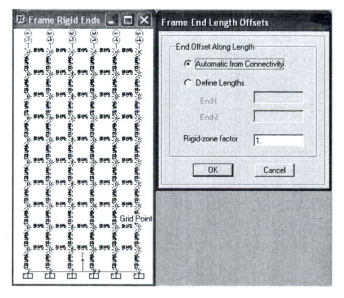

IMAGE 6.30 Assigning End offset, step 1.

Step 22: Assign end offsets, continued. Once the end offsets are created, you will note dark lines at the beam-column joints in the model. (Images 6.31 and 6.32)

IMAGE 6.31 Assigning End offset, step 2.

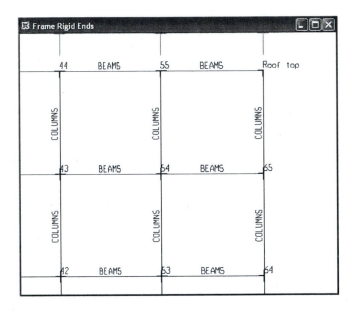

IMAGE 6.32 Assigning End offset, step 3.

Step 23: Assign tensile hinges to the beams. Select the beams. Go to Menu – Assign – Frame/Cable/Tendon – Hinges. (Images 6.33 and 6.34)

Step 24: Assign compression hinges to columns. Select the columns. Go to Menu – Assign – Frame/Cable/Tendon – Hinges. (Image 6.35)

Step 25: Define load cases. Go to Menu – Define – Load Cases. (Image 6.36)

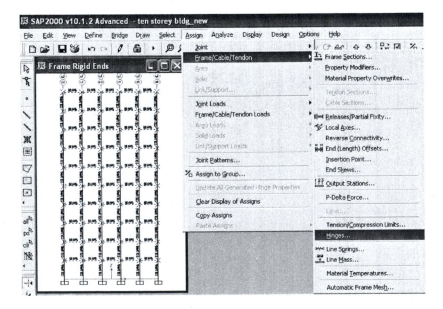

IMAGE 6.33 Assigning tensile hinges to beams, step 1.

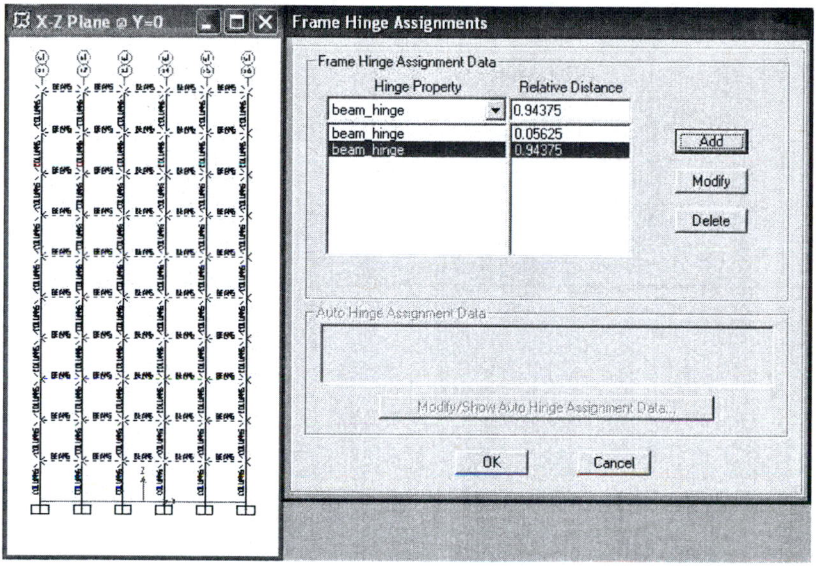

IMAGE 6.34 Assigning tensile hinges to beams, step 2.

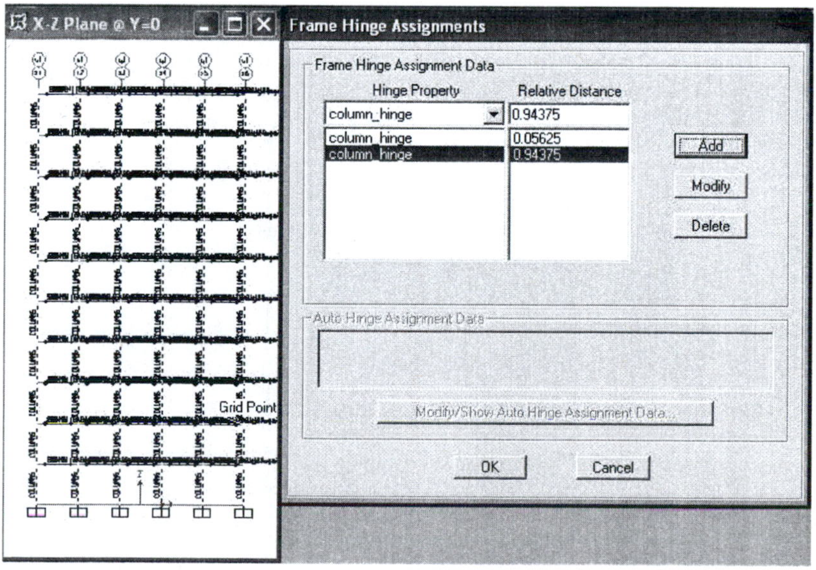

IMAGE 6.35 Assigning compression hinges to columns.

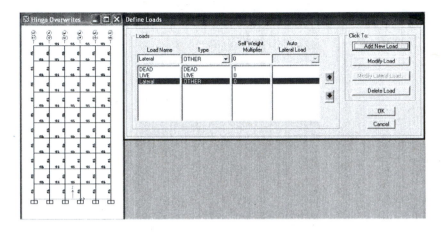

IMAGE 6.36 Defining load cases.

Step 26: Assign the loads to the beam. In this example, we assign a central concentrated load of 41.25 kN, calculated as per the code (IS 1893, 2002), to beams in each floor. Select all the beams. Go to Menu – Assign – Frame/ Cable/Tendon – Point. (Images 6.37 to 6.39)

Step 27: Assign diaphragm action to the model. Select each floor, separately. Go to Menu – Assign – Joint – Constraints. (Images 6.40 and 6.41)

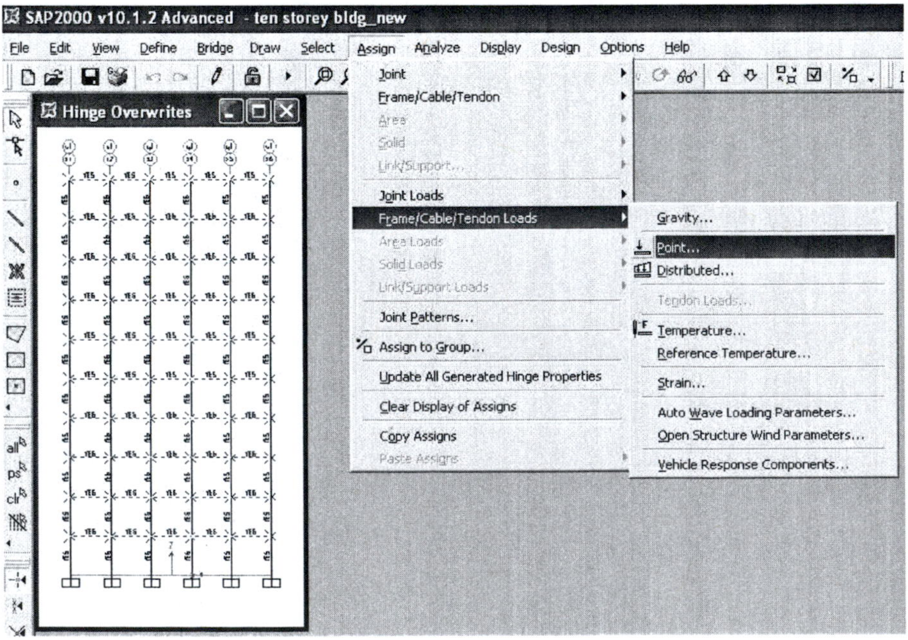

IMAGE 6.37 Assigning loads to beams, step 1.

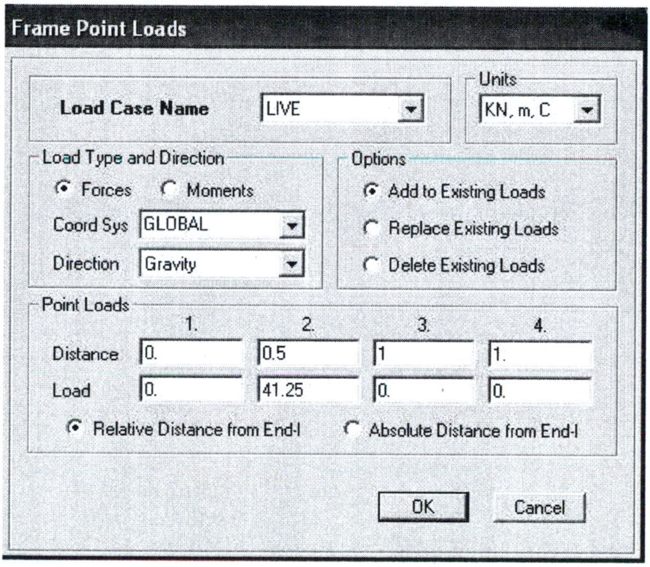

IMAGE 6.38 Assigning loads to beams, step 2.

Step 28: Assign pushover load at the roof top. Select the left top joint (as shown in the screen). Go to Menu – Assign – Joint Loads. Assign 10 kN load in global X direction. (Image 6.42)

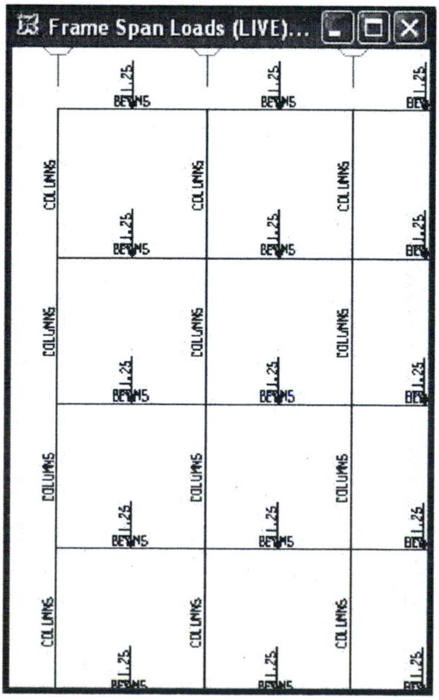

IMAGE 6.39 Assigning loads to beams, step 3.

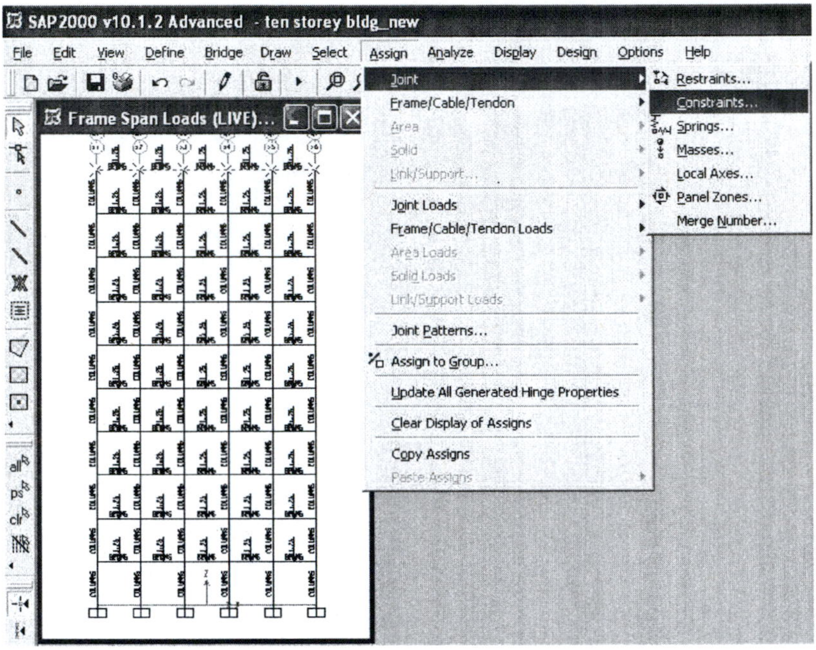

IMAGE 6.40 Assigning diaphragm action, step 1.

Step 29: Define analysis cases. Go to Menu – Define – Analysis Cases – Add New Case. Let the case name be initial pushover. It is essential to apply the dead and live loads on the frame before we intend to push the frame using lateral load. Do not apply pushover load before applying gravity loads. The results could be erroneous. (Image 6.43)

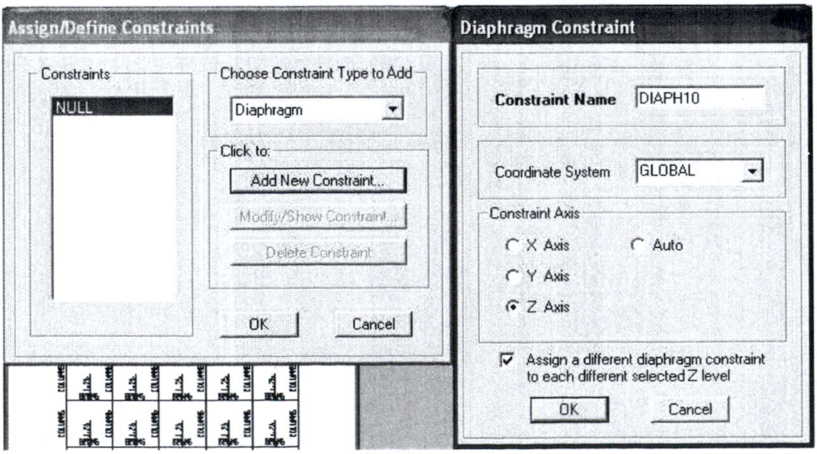

IMAGE 6.41 Assigning diaphragm action, step 2.

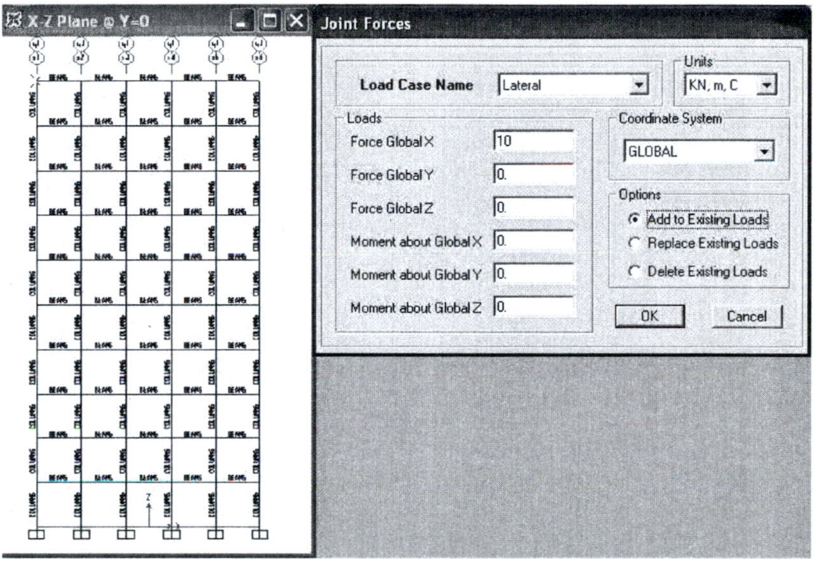

IMAGE 6.42 Assigning pushover load at *roof top*.

Step 30: Define analysis cases, continued. While selecting other parameters, click Load Application – Modify/Show and set the parameters as Full Load. Use monitored displacement at joint *roof top* created in Step 5. (Image 6.44)

Step 31: Define analysis cases, continued. Set a new case—pushover by the same procedure. But we will apply this load case, using displacement control, continuing from the previous case. While defining the parameters, results are saved at multiple states to trace the formation of hinges. (Images 6.45 and 6.46)

Step 32: Define analysis cases, continued. Set the nonlinear parameters. Click Modify/Show. An example case is shown in Image 6.47. However, these parameters are system dependent, and the user can choose these values depending upon the nature of the model, by trial and error (and with experience).

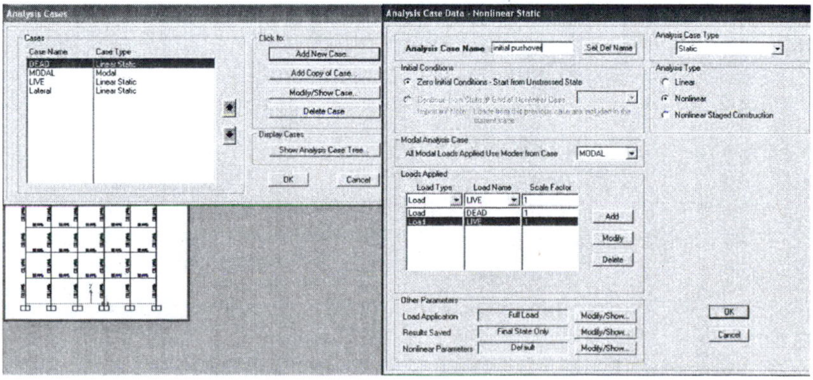

IMAGE 6.43 Defining analysis cases, step 1.

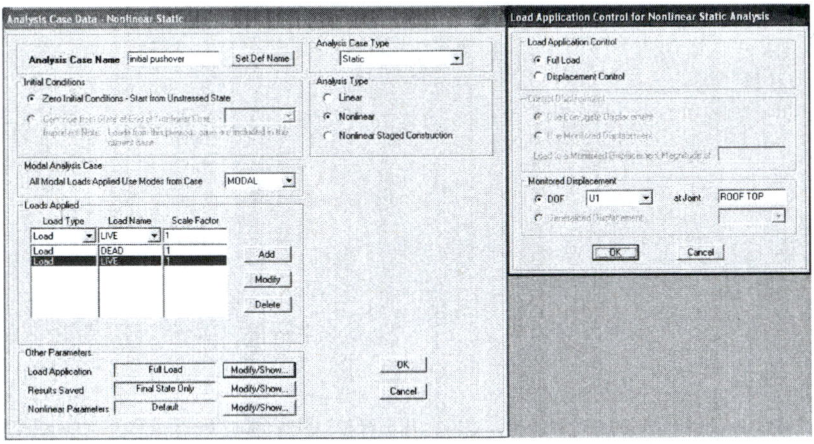

IMAGE 6.44 Defining analysis cases, step 2.

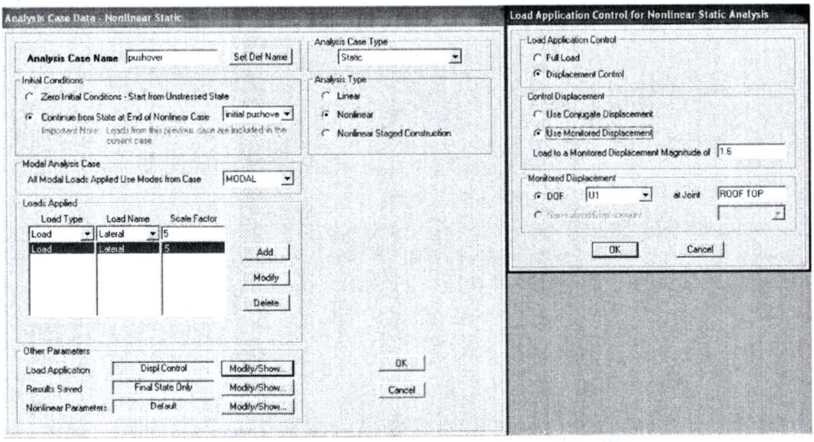

IMAGE 6.45 Defining analysis cases, step 3.

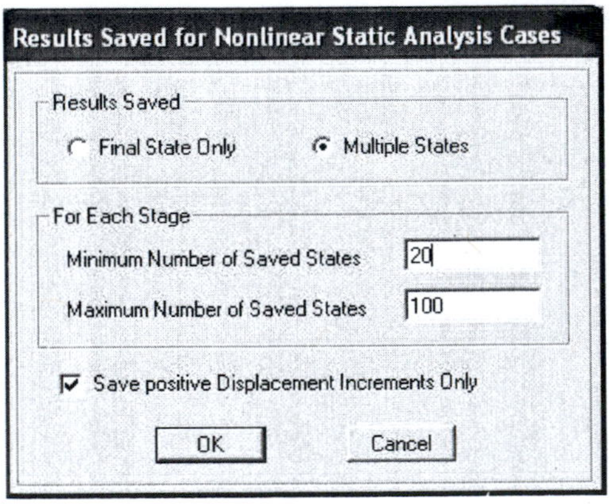

IMAGE 6.46 Defining analysis cases, step 4.

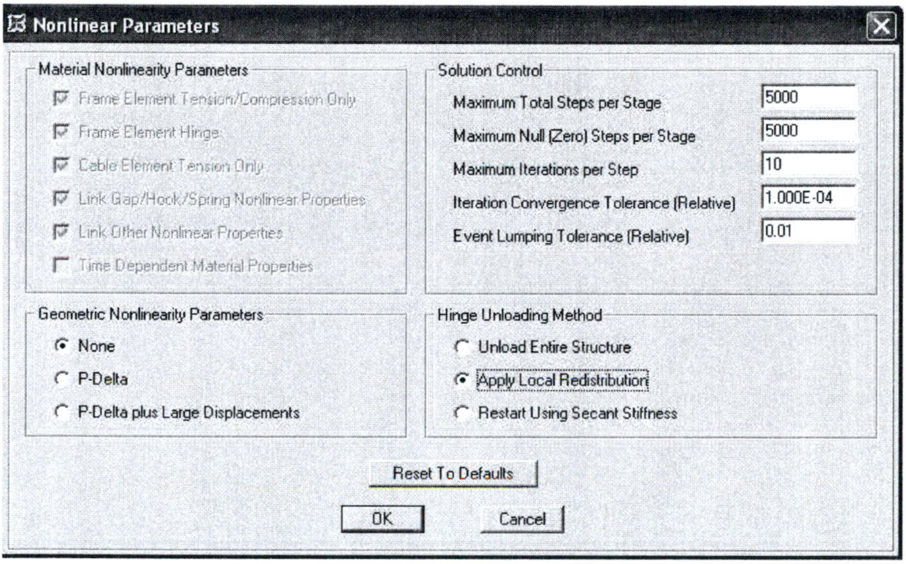

IMAGE 6.47 Defining analysis cases, step 5.

Step 33: Define analysis cases, continued. All defined analysis cases may be viewed and checked using Show Analysis Case Tree. (Image 6.48)

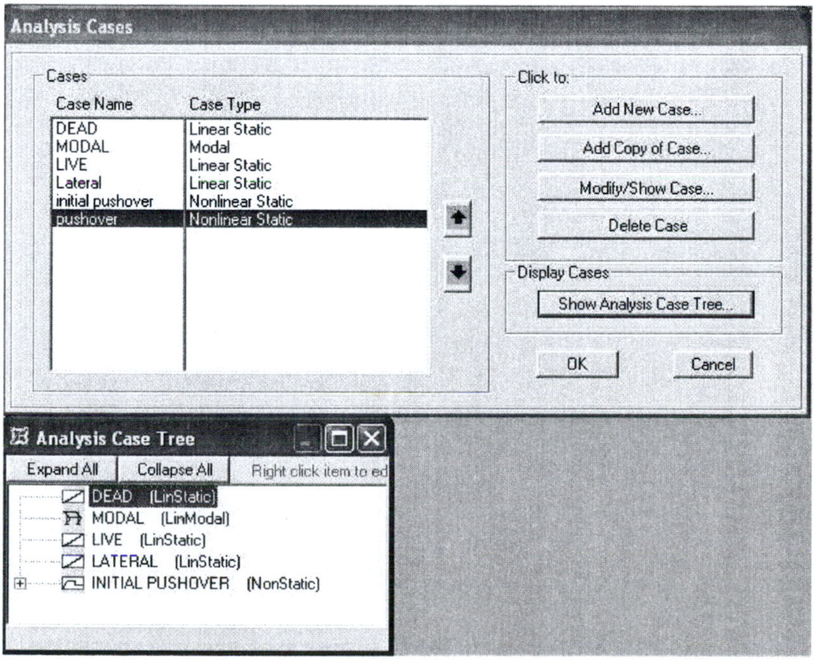

IMAGE 6.48 Defining analysis cases, step 6.

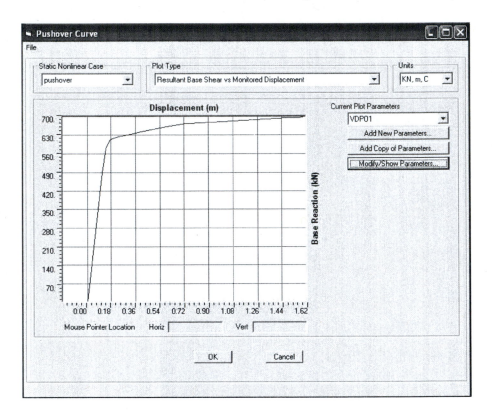

IMAGE 6.49 Pushover curve.

Step 34: Run the analysis. Press F5. To see the pushover curve for a target displacement set at the point *roof tip*, go to Menu – Display – Pushover Curve. (Image 6.49)

Step 35: Obtain the history of formation of plastic hinges. Go to File in the pushover curve screen – Display Tables. (Image 6.50)

Important Note: This example is only a sample illustration to introduce pushover analysis to new users. Although the above steps are believed to introduce this nonlinear static analysis procedure clearly, interpretation of results for any specific model, for any specific purpose, is not the responsibility of the authors. Readers are advised to go through research papers and the Help menu of SAP2000, in detail, for a thorough understanding of different analysis parameters, in their own interest. The aforementioned are only introductory guidelines and solely the interpretation of the software parameters by the authors.

Table Display

File   Edit

Pushover Curve - pushover

Step	Displacemer m	BaseForce KN	AtoB	BtoIO	IOtoLS	LStoCP	CPtoC	CtoD	DtoE	BeyondE	Total
0	-1.792E-17	0.000	220	0	0	0	0	0	0	0	220
1	0.080000	350.293	220	0	0	0	0	0	0	0	220
2	0.111684	489.029	214	6	0	0	0	0	0	0	220
3	0.142984	579.587	172	22	0	0	0	0	26	0	220
4	0.162086	604.419	141	24	22	0	0	0	33	0	220
5	0.178262	613.897	137	8	20	0	0	0	55	0	220
6	0.234813	624.423	131	6	1	0	0	0	82	0	220
7	0.322648	634.455	130	0	0	0	0	0	90	0	220
8	0.402648	642.629	130	0	0	0	0	0	90	0	220
9	0.562235	658.574	126	4	0	0	0	0	90	0	220
10	0.647359	666.804	124	2	0	0	0	0	94	0	220
11	0.729130	673.536	120	2	0	0	0	0	98	0	220
12	0.760624	674.933	120	0	0	0	0	0	100	0	220
13	0.840624	677.060	120	0	0	0	0	0	100	0	220
14	0.920624	679.186	120	0	0	0	0	0	100	0	220
15	1.000624	681.313	120	0	0	0	0	0	100	0	220
16	1.080624	683.440	120	0	0	0	0	0	100	0	220
17	1.160624	685.566	120	0	0	0	0	0	100	0	220
18	1.240624	687.693	120	0	0	0	0	0	100	0	220
19	1.320624	689.820	120	0	0	0	0	0	100	0	220
20	1.400624	691.946	120	0	0	0	0	0	100	0	220
21	1.480624	694.073	120	0	0	0	0	0	100	0	220
22	1.560624	696.200	120	0	0	0	0	0	100	0	220
23	1.600000	697.246	120	0	0	0	0	0	100	0	220

Current Sort String

Current Filter String

Done

IMAGE 6.50 History of formation of plastic hinges.

References

Abu-Lebdeh, T.M. and Voyiadjis, G.Z. 1993. Plasticity-damage model for concrete under cyclic multi-axial loading. *ASCE. J. Eng. Mech.* 119:1465–1484.

Amador, G. and Nadyane, B.-A. 2008. Cumulative ductility spectra for seismic design of ductile structures subjected to long duration motions. Concept and theory background. *J Eq. Eng.* 12(1):152–172.

ATC-40. 1996. *Seismic evaluation and retrofit of concrete buildings.* Redwood City, CA: Applied Technology.

Bangash, M.Y.H. 1989. *Concrete and concrete structures.* Amsterdam: Elsevier Publications.

Candappa, D.C., Sanjayan, J.G., and Setunge, S. 2001. Complete stress-strain curves of high strength concrete. *ASCE. J. Mat. Civil Eng.* 13:209–215.

Carreira, D. and Chu, K.H. 1986. The moment-curvature relationship of RC members. *ACI J.* 83:191–198.

Challamel, N. and Hjiaj, M. 2005. Non-local behaviour of plastic softening beams. *J. Acta Mech.* 178:125–146.

Chandrasekaran, S., Nunzinate, L., Serino, G., and Carannante, F. 2008a. Axial force-bending moment limit domain and flow rule for reinforced concrete elements using Euro Code. *Int. J. Damage Mech.* doi.: 10.1177/1056789 508101200.

Chandrasekaran, S., Nunzinate, L., Serino, G., and Carannante, F. 2008b. Nonlinear seismic analyses of high rise reinforced concrete buildings. *IcFai Uni. J. Struct. Eng.* 1(1):1–7.

Chandrasekaran, S., Nunziante, L., Serino, G., and Carannante, F. 2008c. Axial force-bending moment failure interaction and deformation of reinforced concrete beams using Euro Code. *J. Struct. Eng. SERC* 35(1):16–25.

Chandrasekaran, S., and Roy, A. 2004. Comparison of modal combination rules in seismic analysis of multi-storey RC frames. *In: Proc. 3rd Int. Conf. Vib. Engrg. and Tech (Vetomac),* IIT Kanpur, India: 161–169.

Chandrasekaran, S. and Roy, A. 2006. Seismic evaluation of multi storey RC frames using modal push over analysis. *J. Nonlinear Dyn.* 43(4):329–342.

Chandrasekaran, S., Serino, G., and Gupta, V. 2008. Performance assessment of buildings under seismic loads. *In: Proc. 10th Int. Conf. Struct. under Shock and Impact Loads (SUSI),* Algarve, Portugal: 313–322.

Chandrasekaran, S., Tripati, U.K., and Srivastav, M. 2003. Study of plan irregularity effects and seismic vulnerability of moment resisting RC framed structures. *In: Proc. 5th Asia-Pacific Conf. Shock and Impact Loads,* Changsa, China: 125–136.

Chao, H.H., Yungting, A.T. and Ruo, Y.H. 2006. Nonlinear pushover analysis of infilled concrete frames. *J. Eq. Eng. and Eng. Vib.* 5(2):245–255.

Chen, A.C.T. and Chen, W.F. 1975. Constitutive relations for concrete. *ASCE J. Eng. Mech.* 101:465–481.

Chen, W.F. 1994a. *Constitutive equations for engineering materials, vol. 1: Elasticity and modelling.* Amsterdam: Elsevier Publications.

Chen, W.F. 1994b. *Constitutive equations for engineering materials, vol. 2: Plasticity and modelling.* Amsterdam: Elsevier Publications.

Chopra, A.K. 2003. *Dynamics of structures: Theory and applications to earthquake engineering.* 2nd ed. Singapore: Pearson Education.

Chopra, A.K. and Goel, R.K. 2000. Evaluation of NSP to estimate seismic deformation: SDF system. *J. Struct. Eng.* 126(4):482–490.

D.M. 2005, 14 Settembre. *Norme tecniche per le Costruzioni.* Rome, Italy (in Italian).

D.M. 9 gennaio. 1996. *Norme tecniche per il calcolo, l'esecuzione ed il collaudo delle strut-ture in cemento armato normale e precompresso e per le strutture metalliche.* Rome, Italy (in Italian).

Esra, M.G. and Gulay, A. 2005. A study on seismic behaviour of retrofitted building based on nonlinear static and dynamic analysis. *J. Eq. Eng. Eng. Vib.* 4(1):173–180.

Eurocode: UNI ENV 1991-1. *Eurocodice 1. 1991. Basi di calcolo ed azioni sulle strutture. Parte 1: Basi di calcolo.* (in Italian)

Eurocode: UNI ENV 1991-2. *Eurocodice 1. 1991. Basi di calcolo ed azioni sulle strutture. Parte 2-1: Azioni sulle strutture—Massa volumica, pesi propri e carichi imposti* (in Italian).

Fan, Sau-Cheong and Wang, F. 2002. A new strength criterion for concrete. *ACI J. Struct.* 99:317–326.

FEMA 440. 2005. *Improvements of nonlinear static seismic analysis procedures.* Washington, D.C.: FEMA.

FEMA 450. 2004. *NEHRP recommended provisions and commentary for seismic regulations for new buildings and other structures.* Building Seismic Safety Council (BSSC), Washington, D.C.: FEMA.

Ganzerli, S., Pantelides, C.P., and Reaveley, L.D. 2000. Performance based design using struc-tural optimization. *J. Eq. Eng. Struct. Dyn.* 29:1677–1690.

Ghobarah, A. 2001. Performance based design in earthquake engineering: state of develop-ment. *J. Eng. Struct.* 23:878–884.

Gilbert, R.I. and Smith, S.T. 2006. Strain localization and its impact on the ductility of RC slabs containing welded wire reinforcement. *J. Adv. Struct. Eng.* 9(1):117–127.

Hognestad, E., Hanson, N.W., and McHenry, D. 1955. Concrete stress distribution in ultimate strength design. *ACI J.* 52(6):455–479.

Hsieh, S.S., Ting, E.C., and Chen, W.F. 1982. A plastic-fracture model for concrete. *J. Solids Struct.* 18:181–197.

IS 456. 2000. *Plain and reinforced concrete: Code of practice. Fourth revision.* New Delhi, India: Bureau of Indian Standards.

IS 1893. 2002. *Criteria for earthquake resistant design of structures: Part 1. General provi-sions for buildings, fifth revision.* New Delhi, India: Bureau of Indian Standards.

IS 13920. 1993. *Ductile detailing for reinforced concrete structures subjected to seismic forces.* New Delhi, India: Bureau of Indian Standards.

Jirasek, M. and Bazant, Z.P. 2002. *Inelastic analysis of structures.* New York: Wiley.

Khan, A.R., Al-Gadhib, A.H., and Baluch, M.H. 2007. Elasto-damage model for high strength concrete subjected to multi-axial loading. *Int. J. Damage Mech.* 16(3):367–398.

Ko, M.Y., Kim, S.W., and Kim, J.K. 2001. Experimental study on the plastic rotation capacity of reinforced high strength beams. *J. Mat. Struct.* 34:302–311.

Lopes, S.M.R. and Bernardo, L.F.A. 2003. Plastic rotation capacity of high-strength concrete beams. *J. Mat. Struct.* 36:22–31.

Mahin, S., Anderson, E., Espinoza, A., Jeong, H., and Sakai, J. 2006. Sustainable design con-siderations in the construction to resist the effects of strong earthquakes. *In: Proc. 4th Int. Workshop Seismic Design and Retrofit of Transportation Facilities.* Burlingame, CA: 10.

Menetrey, P.H. and Willam, K.J. 1995. Tri-axial failure criterion for concrete and its general-ization. *ACI J. Struct.* 92:311–318.

Mo, Y.L. 1992. Investigation of reinforced concrete frame behaviour: Theory and tests. *Mag. Conc. Res.* 44(160):163–173.

Nunziante, L., Gambarotta, L., and Tralli, A. 2007. *Scienza della Costruzioni.* Milan: McGraw-Hill (in Italian).

Nunziante, L. and Ocone, R. 1988. *Limit design of frames subjected to seismic loads.* Napoi, Italy: CUEN Publications (Cooperativa Universitaria Editrice Napoletana).

Ordinanza 3316: Correzioni e modifiche all'ordinanza 3274. 2005. *Modifiche ed integrazioni all'ordinanza del Presidente del Consiglio dei Ministri n. 3274 del 20 Marzo 2003.* Rome, Italy (in Italian).

Ordinanza del presidente del consiglio dei ministri del 20 marzo. 2003. *Primi elementi in materia di criteri generali per la classificazione sismica del territorio nazionale e di normative tecniche per le costruzioni in zona sismica.* Rome, Italy (in Italian).

Ottosen, N.S. 1977. A failure criterion for concrete, *ASCE. J. Eng Mech.* 103:527–535.

Pankaj, A. and Manish, S. 2006. *Earthquake resistant design of structures.* New Delhi, India: Prentice Hall of India Pvt. Ltd.

Papadrakakis, M., Fragiadakis, M., and Lagaros, N.D. 2007. *Extreme man-made and natural hazards in dynamics of structures.* Dordrect: Springer.

Park, H. and Kim, H. 2003. Microplane model for reinforced concrete planar members in tension-compression. *ASCE J. Struct. Eng.* 129:337–345.

Park, R. and Paulay, T. 1975. *Reinforced concrete structures.* New York: John Wiley and Sons.

Paulay, T. and Priestley, M.J.N. 1992. *Seismic design of RC and masonry buildings.* New York: John Wiley and Sons.

Pfrang, E.O., Siess, C.P., and Sozen, M.A. 1964. Load-moment-curvature characteristics of RC cross-sections. *ACI J.* 61(7):763–778.

Pisanty, A. and Regan, P.E. 1993. Redistribution of moments and the possible demand for ductility. *CEB Bulltn d' Information* 218:149–162.

Pisanty, A. and Regan, P.E. 1998. Ductility requirements for redistribution of moments in RC elements and a possible size effect. *J. Mat. Struct.* 31:530–535.

Priestley, M.J.N., Calvi, G.M., and Kowalsky, M.J. 2007. *Displacement based seismic design of structures,* Pavia, Italy: IUSS Press.

Raphel, F., Marak, V.W., and Truszcynski, M. 2002. *Constraint LINGO: A program for solving logic puzzles and other tabular constraint problems.* Berlin: Springer.

Rustem, S. 2006. Structural damage evaluation. *J. Jpn. Assoc. Eq. Eng.* 4(2):39–47.

Sankarasubramaniam, G. and Rajasekaran, S. 1996. Constitutive modelling of concrete using a new failure criterion. *J. Comp. and Struct.* 58:1003–1014.

SAP2000 Advanced Version: 10.1.2. Structural Analysis Program, Computers and Structures Inc., Berkeley, CA (http://www.csiberkeley.com).

SEAOC. 1995. *Performance based seismic engineering of buildings: Vision 2000 report.* California, USA: Structural Engineers Association of California.

Sforza, A. 2002. *Modelli e metodi della Ricerca Operativa.* Roma: Edizioni Scientifiche Italiane (In Italian).

Sinan, A. and Asli, M. 2007. Assessment of improved nonlinear static procedures in FEMA 440. *ASCE J. Struct. Eng.* 133(9):1237–1246.

Student LINGO. 2005. Parameter optimization software. Release 9.0: LINGO Sys Inc.

Sumarec, D., Sekulovic, M., and Krajcinovic, D. 2003. Failure of RC beams subjected to three point bending. *Int. J. Damage Mech.* 12(1):31–44.

Wolfram Mathematica Software. Ver 6.0.0. for Windows Platform: Wolfram Research Inc.

Wood, R.H. 1968. Some controversial and curious developments in plastic theory of structures. *In: Engineering plasticity* (J. Heyman, FA. Leckie eds). Cambridge, UK: Cambridge University Press, pp. 665–691.

Yakut A., Yilmaz, N., and Bayili, S. 2001. Analytical assessment of seismic capacity of RC framed building. *J. Jpn. Assoc. Eq. Eng.* 6(2):67–75.

Zhang, Y. and Der Kiureghian, A. 1993. Dynamic response sensitivity of inelastic structures. *J. Comp. Methods Appl. Mech. Eng.* 108:23–36.

Index